T0269161

CAMBRIDGE LIBRARY COLLECTION

Books of enduring scholarly value

Life Sciences

Until the nineteenth century, the various subjects now known as the life sciences were regarded either as arcane studies which had little impact on ordinary daily life, or as a genteel hobby for the leisured classes. The increasing academic rigour and systematisation brought to the study of botany, zoology and other disciplines, and their adoption in university curricula, are reflected in the books reissued in this series.

Land and Sea

Philip Henry Gosse (1810–88) is best remembered today for the portrait given by his son Edmund in his autobiographical *Father and Son*. In his own day, he was famous as a natural historian, and his books were extremely popular. (His *Naturalist's Sojourn in Jamaica* is also reissued in this series.) In 1857, Gosse moved from London to Devon, where he spent the rest of his life. This 1865 book offers essays about various aspects of the geography and natural history of the West Country. There are some digressions (one chapter is on the woods of Jamaica), and reminders of the two great Victorian crazes, for ferns and for seashore life, which Gosse's writings partly instigated. In his final essay, on Dartmoor, is an appendix which argues that Britain is the biblical Tarshish – a reminder that Gosse was also a fundamentalist Christian who struggled with many aspects of contemporary science.

Cambridge University Press has long been a pioneer in the reissuing of out-of-print titles from its own backlist, producing digital reprints of books that are still sought after by scholars and students but could not be reprinted economically using traditional technology. The Cambridge Library Collection extends this activity to a wider range of books which are still of importance to researchers and professionals, either for the source material they contain, or as landmarks in the history of their academic discipline.

Drawing from the world-renowned collections in the Cambridge University Library and other partner libraries, and guided by the advice of experts in each subject area, Cambridge University Press is using state-of-the-art scanning machines in its own Printing House to capture the content of each book selected for inclusion. The files are processed to give a consistently clear, crisp image, and the books finished to the high quality standard for which the Press is recognised around the world. The latest print-on-demand technology ensures that the books will remain available indefinitely, and that orders for single or multiple copies can quickly be supplied.

The Cambridge Library Collection brings back to life books of enduring scholarly value (including out-of-copyright works originally issued by other publishers) across a wide range of disciplines in the humanities and social sciences and in science and technology.

Land and Sea

Philip Henry Gosse

CAMBRIDGE
UNIVERSITY PRESS

CAMBRIDGE
UNIVERSITY PRESS

University Printing House, Cambridge, CB2 8BS, United Kingdom

Cambridge University Press is part of the University of Cambridge.
It furthers the University's mission by disseminating knowledge in the pursuit of
education, learning and research at the highest international levels of excellence.

www.cambridge.org
Information on this title: www.cambridge.org/9781108073424

This edition first published 1865
This digitally printed version 2014

ISBN 978-1-108-07342-4 Paperback

LAND AND SEA.

BY

PHILIP HENRY GOSSE, F.R.S.

LONDON:

JAMES NISBET & CO., 21 BERNERS STREET.

M.DCCC.LXV.

EDINBURGH :
PRINTED BY BALLANTYNE AND COMPANY,
PAUL'S WORK.

PREFACE.

THIS volume has no particular *subject*. Its contents pretend to no connexion, no continuity: each of the Papers must be read as perfectly isolated from all the rest. Several of them, indeed, are pen-pictures of associated scenes, and of kindred things that have occurred in them. Others touch of kindred things that occur elsewhere. The whole may be likened to a handful of sketches taken at random out of one artist's portfolio.

TORQUAY, *December* 1864.

CONTENTS.

LUNDY ISLAND.

LUNDY ISLAND—(continued.)

LUNDY ISLAND—(continued.)

CONTENTS.

LIST OF ILLUSTRATIONS.

Lundy Island.

LUNDY ISLAND.

THE EASTERN COAST (*from the Landing.*)

THERE are many odd nooks and corners in England which are seldom visited by tourists, and of which topographical writers know next to nothing, which are yet well stored with objects of interest amply sufficient to repay the toil and ingenuity expended in searching them out. Such a spot is Lundy, that little rocky island with precipitous sides that stands in the midst of the waters of the Bristol Channel, like a sentinel, to guard this great sea-road into the heart of England. I had been prosecuting some researches among the microscopic zoophytes, and other objects of marine natural history, on the picturesque coast of North Devon, through early summer; and from the lofty downs and cliffs around Ilfracombe I had often gazed out upon Lundy, a long low wall of purple in the horizon, and wished to explore it. It can be seen only in the clearest weather: many a day I have looked for it in vain; and thus its appearance became associated with lovely mornings and clear golden sunsets; and what I had heard of some peculiarities in its zoology, and what I imagined an insular rock so situated might afford to the naturalist, determined me to take the earliest opportunity of a visit to its cliffs.

Such an occasion was found through the courtesy of Hudson Heaven, Esq., the eldest son of the proprietor of the island, who kindly invited myself and two companions to accompany him in his boat, about to sail. Accordingly, the break of day on the 1st of July saw us on the little quay at Ilfracombe, with portmanteaus and carpet-bags, collecting-basket, bottles and jars for zoophytes, and some packets of sandwiches and other comforts for the interior organisation. We had to wait at least an hour after the time appointed, before the tide served: it was rather a cold morning; the sky was leaden, and there was already a tough breeze from the westward, dead against our course, which seemed likely to freshen: the fishermen, moreover, that sauntered out from their hovels at that early hour, assured us, to keep up our already wavering courage, that there was a pretty heavy sea running outside. However, we were booked for the voyage, and were not going to retreat because it might have a dash of adventure: indeed, the heroism of one of the party was so strung up by the exciting prospect, that he boldly intimated his purpose of joining the search for Franklin, after this expedition.

So at length we stowed ourselves in the stern-sheets; the peak was hoisted, the jib was set, the mainsail trimmed; another pull upon the peak-halyards, the jib and main sheets tautened, and here we were with the red sails as flat as a pancake, facing the westerly breeze, and pitching and rolling in the wash of the sea, which is always more than ordinarily uproarious off the harbour's mouth just at the turn of the tide.

The little boat ploughed and dug through the green and foaming waves, quivering now and then as one struck her broadside in a way that rather put a damper upon our mirth. Before she had made one short tack, and before we were well abreast of the flag-staff that crowns Capstone Hill, an envious sea curled up its green head right over the quarter, and broke upon us, drenching us as completely as if we had invaded its domain instead of its intruding into ours. A pretty pickle this to begin an eight hours' voyage with! and very comforting to the stomachs, already receiving awful warnings of what was about to be. We all grew as mute as mice in no time:

the enthusiasm of science, no more than the pleasure
of holiday-making, can bear up with dignity against
the manifold inflictions of cold and wet, cramped
limbs, and the perpetual eversion of that internal
organisation I spoke of just now, which no sandwiches
could soothe. But let that pass.

The approach to the island was interesting; especially as our kind cicerone, Mr Heaven, pointed out
the different objects of interest, and gave us legendary
and statistical information. Its form somewhat resembles that of an oak-leaf, being considerably sinuated in outline; and the narrow peninsula of Lametry,
constituting its southern extremity, with Rat Island
as its termination, we may call the footstalk of the
leaf. This end of the island curves round to the eastward, partially enclosing a little bay with good anchorage, pretty well sheltered from all but easterly winds.
About twenty vessels were lying here at anchor, of
various nations and of all sizes, from the stately three-
masted ship to the tiny fishing skiff. On our expressing surprise at seeing so many craft, Mr Heaven
assured us that often there were many more. " I
have known," said he, " three hundred vessels in sight
at once. On one occasion the wind had hung long
from the westward, and had kept-in the outward-
bound craft: it at last changed and allowed them to
sail, but suddenly shifting again, and coming on to
blow heavily from the old quarter, a hundred and
seventy vessels put back and anchored in our little
roadstead, all vessels of size, not counting boats."

The only landing-place on the whole island is in
this bay; and here the Trinity House have made a
good carriage-road from the beach up the precipitous
hill-side to the lighthouse, which occupies the highest
point, and which I shall speak of more particularly
presently. Up this zigzag road, which is substan-
tially built with granite in the lower part, where it is
exposed to the action of the sea in heavy gales, we
climbed, eager to find the means of satisfying our
quickened appetites, yet not indifferent to the charms
with which nature had embellished this lonely place.
The sides of the road were gay with flowers of many
kinds. The common mallow, the milfoil, the weld or
wild mignonette, looking like its pleasant namesake,
but scentless; the flaring ox-eye daisy, the figwort,
with its brown bead-like blossoms; the navew, loose
and sprawling, but bright in hue; ragworts and sow-
thistles, and elder-bushes with snow-balls of bloom,
the nearest approach to a tree which the island can
boast; these, with minor weeds and grasses and ferns
of several kinds, fringed the footpath. The perpendicular side of the road, where the shale had been
scarped away, and the crevices of the stones, where it
had been faced with a rude wall, presented other and
more attractive features. The kidney vetch, or lady's
finger, displayed its heads of delicate flowers in profusion, pale yellow fading into cream colour; and the
scarlet-tipped blossoms of the little bird's-foot lotus,
that characteristic plant of our seaward downs and
precipitous slopes, were not less abundant. From

between the loose stones the navelwort shot out its
singular spikes, each springing perpendicularly from
a bed of succulent shield-like leaves, and fringed to
its tall summit with little drooping bells of yellowish
white.   The situation seems particularly agreeable to
this plant, for we found it in many parts of the island
growing in great luxuriance, some of the spikes eighteen or twenty inches in height, and thickly covered
with flowers.   The herb Robert, the bitter vetch, and
the purple sandwort, displayed their unobtrusive but
pretty blossoms among the herbage; and the crimson
bells of the common heath, already opened, were
fringing the edges of the slope above our heads.   The
sheep's-bit scabious, a lovely flower, with globose
heads of azure blue, was not wanting; and the surface of the rock was covered here and there with
broad patches of the white stone-crop, whose white, or
rather carnation-coloured, starry blossoms were conspicuously beautiful.   But more prominent than all
was that noblest of British flowers, the tall foxglove,
flourishing in special luxuriance and beauty, while
fragrance was diffused from scores of honeysuckles
that climbed and sprawled on every side.

All these and other plants, some greeting us as old
acquaintances, others possessing the charm of comparative novelty, were an agreeable contrast to the
desolation and barrenness we had pictured to ourselves as reigning here.   And as we proceeded we
saw pleasant traces of feminine taste, for gentle hands
had been busy in sowing seeds of stocks, and wall-

flowers, and nasturtiums in the nooks of the rock,
which were now beginning to spread the beauty of
their foliage over the ruggedness, and gave promise
of additional beauty by and by.

The island is the property of William Heaven,
Esq., who has erected a handsome mansion above the
landing-place, in a sheltered hollow, which commands
an extensive view of the opposite coast of Devon and
of the broad Bristol Channel. Here he resides with
his amiable family, exercising a patriarchal rule over
his little dominion. Two thousand acres form his
realm; of which a considerable portion is under cultivation, and is let to a tenant farmer, John Lee by
name, familiarly known as Captain Jack, an excellent, worthy man. In his earlier days he was bred to
the sea, but now he ploughs the land. At his house,
" The Farm," visitors are entertained ; we found
accommodations decent, (for the circumstances), a
well-supplied table, attendance prompt and kindly,
and charges moderate. With the exception of the
lighthouse-keeper, who with his family and subordi-
nates occupies a substantial stone house at the foot of
the lighthouse, on the western edge of the island,
rather remote from the Farm, the rest of the inhabitants are labourers, and their families employed in
husbandry, or in the mechanical occupations that
minister to it.

The whole population amounts to about fifty souls,
not one of whom is a native of the isle: a child has
not been born here within the memory of the present

generation; the women invariably going over to the mainland when their confinement approaches. No medical man resides on the island; but a fire lighted on a particular summit summons a boat in cases of emergency, from the little village of Clovelly, just opposite. This place, itself a spot of romantic beauty, one of the gems of the North Devon coast, is situated in Barnstaple Bay, just within Hartland Point, (the Herculis Promontorium of Ptolemy), and is distant about five leagues from the end of Lundy. A boat comes across every Friday, bringing the week's accumulations of the post-office, and returns with any letters that are ready. Other communication with the shore is only casual, as when the Pill boats come down as far as this from their little pilot village at the Avon's mouth to look out for ships, and anchor in the bay; or when a skiff-load of lobsters is run up to Ilfracombe to be shipped, per steamer, for Bristol.

A mutton-chop, improvised by Captain Jack's larder, revived our vigour, and we sallied out towards the south end to reconnoitre. A walk between stone fences, enlivened by many interesting plants in flower, some of which I shall mention presently, led us to the ruins of the castle, bearing the name of the De Mariscos, the earliest possessors of the island on record, who held it as long ago as Henry the Second's time. A legend is told of one of this family, illustrative of the bold lawlessness of the times, as well as of the natural strength of this island. It rests on the authority of the contemporary historians.

In the year 1238, William de Marisco conspired
with a knight of the palace to murder King Henry
III. The act of assassination was entrusted to the
courtier, who gained access to the royal chamber by
climbing up to the window. It chanced, however,
that the king lay elsewhere that night, and the con-
spirator, thus baffled, sought his victim in other
chambers. Ignorant whither to go, he at length
bursts into an apartment with his dagger drawn in
his hand, where sits a lady of the court reading. An
alarm is instantly raised, the servants crowd in, and
the villain is taken. He, poor wretch, expiated his
intended crime by being drawn asunder at Coventry
by four horses; while his coadjutor, Marisco, fled to
his island of Lundy, strengthened his castle, fortified
the accessible parts of the cliffs, and became a pirate.
For some years he did much damage, ravaging the
neighbouring coast, and intercepting ships; but at
length, being surprised by the king's forces, he, too,
suffered death.

The walls of the castle and the ancient keep remain
in integrity, and have been turned, by the addition of
new walls, into labourers' cottages, the chimneys of
which peep out from the ruins, so as greatly to mar
their picturesque effect.

A woman was standing at one of the doors, and
children were playing round: we shuddered to see
the little things run and jump on the edges of the
precipice, and babies carry babies a little younger than
themselves into places where a single false step would

have plunged them fathoms down; and we spoke to
the good woman about the danger. Such, however,
is the power of habit to create indifference, that she
actually appeared not to understand what was meant.
Great mixens outside the doors, strewn with the shells
of enormous limpets, and with those of the green
conical eggs of guillemots, afforded amusing evidence
of the favourite food of the poorer inhabitants of the
island.

A few rods below the castle, where the greensward
slopes steeply down to the south-east, a sort of doorway in the hill-side attracted our notice, and we
looked in. It was the entrance to a large chamber
excavated out of the solid rock, and bore indubitable
proofs of its being a work of art. The gray shale of
which this end of the isle is composed is friable, and
easily removed; and time and labour alone would be
needed to form such a cavern as this. A long slab,
resting on two upright ones for joints, made the doorway. The cave is now used as an occasional stable,
but tradition assigns a very different purpose for its
construction. It is called Benson's Cave, and its
history is as follows:—

" Exactly a century ago, the member of Parliament
for Barnstaple was one Thomas Benson, a man of
more talent than character. He was the owner of a
ship called the *Nightingale*, which having been lost
on her outward voyage to Maryland, he claimed the
insurance. Before it was paid, however, one of the
crew of the sunken ship gave information which led

to the exposure of an artfully-planned piece of villany.
It was proved that Benson, having shipped a valuable
cargo of linen and pewter, with a ballast of salt, gave
secret orders to the master to remain off Lundy,
whither he repaired. The crew were here tampered
with, and, by bribes and threats, were induced to
comply with the proposed scheme. The linen and
pewter were landed and concealed in this new-made
cavern, excavated by Benson for the express purpose.
The ship then sailed; but meeting in the mouth of
the channel a homeward-bound vessel, the master
thought it a good opportunity to execute his purpose.
He went below, bored a hole through the bottom,
and knocked down the bulk-heads, that the water
might get at the salt. But the sea pouring in with
great rapidity, and the strange vessel being yet a
good way off, it was thought they might possibly not
be able to reach her. The mate then fired the oakum
stores with a candle, having first stopped the leak
with a marling-spike. The smoke and flame were
soon seen on board the approaching ship, which presently bore down, and, taking the crew on board,
carried them into Clovelly. Protests were sworn at
Bideford; but meanwhile the boatswain, conscience-
stricken, gave information of the roguery. The arch-
villain Benson escaped to Portugal: his subordinate,
the master, Lancey, was hanged ; and the cavern
remains to this day to perpetuate the remembrance
of their crimes."

The steep sunny slopes of this part of the island

were gay with the purple bloom of the cinereous heath, and with the brilliant masses of blossom of the yellow broom. A bush of this latter kind was springing out of the very lintel of the cavern doorway, and its long spikes of flowers were elegantly pendent over the entrance, the darkness of the interior throwing out into fine relief the rich golden mass of bloom. The thorny, or Burnet-leaved rose, was trailing its lengthened and tortuous branches over the ground, nowhere rising to more than a few inches in height: we were charmed with the beauty and delicacy of its spotless cream-coloured blossoms, and still more with their exquisite fragrance. We afterwards found this plant quite characteristic of the botany of the island.

From these slopes we looked down upon, but did not explore, the peninsula of Lametry, a mass of land precipitous on every side, and joined to the main of the island by a ridge of rock running up to a sharp knife-like edge. Beyond this is an insular rock called Rat Island, from the great number of rats that have made it their home. They are believed to feed largely on fish, as well as on limpets and other littoral prey. Lundy is much infested with rats. For a while the old English, or black rat, succeeded in maintaining undisturbed possession of this little nook against its ruthless exterminator the Norway, or brown rat. The latter, however, has at length found its way across, and is already the more numerous of the two. Mice are quite unknown.

Among the lovelier plants we noticed the little

euphrasy, that tiny flower that derives its name of eyebright, not from its beauty, though few lovers of flowers behold it without brightening eyes, but from its old reputation for "making old eyes young again," a reputation which, if Milton may be believed, is as old as the days of Adam at least, for the Archangel, about to guide our first parents' gaze into distant ages,—

> " The film removed
> Which that false fruit, which promised clearer sight,
> Had bred ; then purged with euphrasy and rue
> The visual nerve, for he had much to see."

The little shining geranium, the dwarf red rattle, the yellow tormentil, and that universal favourite, the scarlet pimpernel, were scattered in the bordering herbage of the paths ; and the walls of uncemented stone were nearly covered with large patches of white stone-crop, and of wild 'thyme, both beautiful but minute plants, the pink blossoms and downy capsules of the latter particularly noticeable from their abundance. Here, also, as well as in other places, grew in great profusion, the wood germander, or bitter sage, whose wrinkled leaves were used during the scarcity of the last war as a substitute for tea.

Fortunately, however, we were not reduced to any such sorry alternative, for our worthy old landlady's tea-caddy proved well stocked with the real China leaf ; and when we got back from our afternoon's stroll, we did justice to its revivifying qualities.

The next morning we started, under the auspices of

our courteous guide, to visit the north end, the resort of countless sea-birds, and in going to it we skirted along the eastern side. Viewed from the road above the landing-place, this line of coast presents a curious appearance.* The gray cliffs rise nearly perpendicularly from the sea, to a height varying from fifty feet to as many yards; then a broad green slope very even and regular, forming an angle of 45° with the horizon, (less or more,) carries up the elevation to four or five hundred feet, and there is the flat summit. The regularity of these slopes is remarkable, and one is ready to fancy that some gigantic carpenter has been at work, bevelling off the edge with a plane. From the sea the deep rich verdure of this inclined surface has a very attractive appearance, and when looked at narrowly, has a roughened texture, like that of a close-grown forest. This is owing to the nature of the herbage, which consists almost exclusively of the common brake-fern. In winter, as we were informed, the brown hues, assumed by this plant in decay, give to this side of the island a russet tint particularly rich and mellow.

One of the first things that attracted our attention, and that continued to excite interest, was the extraordinary abundance of the cocoons of a small species of hawkmoth, known to collectors as the Burnet-moth. In the open waste places, the stalks of grass and the slender stems of herbaceous plants were studded with

* This view is seen on page 3. All the views of the island are from sketches taken by myself on the occasion of this visit.

these little appendages by hundreds, or even by thousands. The cocoon is a pretty object; it is of a spindle shape, that is, swollen in the middle, and pointed at each end. It is formed of silk compacted into a papery substance, bright yellow and glistening, and is attached to the grass perpendicularly all along one side. Some of them which I opened displayed the caterpillar as yet unmetamorphosed, an inert little creature of a pale yellow, studded with rows of close-set black spots. Others contained the black shining chrysalis, in which I detected a curious habit. I had collected a dozen or two stalks with cocoons, and had brought them into my bedroom. At night, while sitting reading, I perceived some faint creaking sounds proceeding from them, and by bringing each in succession close to my ear, I was enabled to find out the individuals from which the noise issued. Then holding the cocoon between the eye and the light, its semi-transparency permitted me to see the enclosed pupa busily engaged in revolving on its long axis, and the sound was caused by the grating of its rings against the papery walls of its prison.

We found multitudes of the moths sitting on the herbage, or flitting hither and thither on feeble wing. Many were drying their half-expanded wings in the morning sun; some were pushing their way out of the upper extremity of the brittle cocoon, previously to bursting the chrysalis skin; and others were emerging from the projected pupa, so wet and shrivelled, that it seemed marvellous that those crumpled and dis-

torted wings should in an hour become the elegant
organs which we afterwards see them, smooth and
satiny, or rather burnished with that rich subdued
gloss that we see in what is called frosted gold, dark,
sea-green, spangled with large spots of crimson.

Truly, in studying so insignificant and lowly a
creature as this, sown broadcast as it were upon the
wild moors of this island rock, we cannot help being
struck with the lavish pains (to speak according to
the manner of men) that have been bestowed upon
it. How elegantly has it been fashioned and trimmed ;
how gorgeously painted and gilded ; how carefully
provided for ! Surely he must be blinder than the
mole who does not trace here

> " The unambiguous footsteps of that God
>   Who gives the lustre to an insect's wing,
>   And wheels His throne upon the rolling worlds."

We wended our way along a narrow path through
the tall fern, occasionally entangled among the tortuous branches of the sweet honeysuckle, or catching
our feet in the trailing shoots of the white rose. How
different the odour of these two flowers ! Both are
sweet, but the fragrance of the rose is far superior as
an aroma to the sugary scent of the honeysuckle.
Tall foxgloves, everywhere springing up from the
dense bed of brake, gave quite a character to the
scene. I think I never saw this magnificent flower
in so fine a condition ; several spikes occurred fully
six feet in height, straight as an arrow, and densely
crowded with their large purple bells. Our friend

assured us that he had counted, on a specimen of
extraordinary dimensions, the remarkable number
of three hundred and sixty-five flowers, exclusive of
unexpanded buds. This must have been a giant.
We could not have selected a more propitious time
for seeing nature in her loveliness; it was what Virgil
elegantly calls "formosissimus annus,"—the year in
the height of beauty. The opening of July is the
season when more plants are in flower than at any
other period; the joyous insects are gay upon the
wing, and those birds that are so inseparably associated with lovely summer weather are all with us;
the atmosphere is apt to be calm and clear, and the
deep transparent azure of the sky is reflected with a
deeper intensity from the sparkling sea, just as we
saw it now, as from our bowery walk we ever and
anon gazed out upon the broad main, the white sails
scattered over its surface, gleaming in the morning
sun, and answering to the fleecy clouds that flitted
over the face of heaven.

> "Land and sea
> Give themselves up to jollity."

Several tiny streamlets ooze out from the upland
moors, and trickling down the sloping sides find their
way along the chines and gullies to the sea. The
spongy nature of the soil, and the matting of the
vegetation impeding the flow of the water, cause the
courses of these streams to form bogs, difficult to pass,
but presenting some objects of interest. In the first
that we came to we found two kinds of speedwell, the

lovely germander, familiar to every one as the blue-
eyed gem of the hedge-bank, and the spiked speedwell, a smaller species and much more rare, and
rather to be looked for in chalky pastures than on the
swampy borders of a stream.   That plant, sacred to
friendship, the true forget-me-not, was also abundant
here, together with a white variety of the same species
that I have not seen noticed.

In another similar brook that breaks out from its
darkling bed beneath dwarf willows, the common
buttercup of our meadows was growing in company
with a much more uncommon species of the same
genus, the great spearwort; the latter we found by
no means rare in various parts of the island.

The dwarf-furze, a smaller kind than that of our
commons and downs, overruns a considerable portion
of the central part of the isle, mingling freely with
the fine and the cross-leaved heaths, and the ling or
true heather : this last was not indeed yet in blossom,
but the true heaths were in full flower.   The white-
blossomed variety of the cross-leaved heath we found
not uncommon, readily distinguished from the ordinary state of the plant, not more by the pure creamy-
white of its bloom, contrasting with the rosy purple
hue which is normal, than by a pale yellow-green
characterising the foliage, by which patches could be
discriminated almost as far as they could be seen.

How delightful it is, when tired with exercise, to
throw one's weary limbs upon the soft yet springy

heather, which yields and yet sustains, with the elasticity of a hair mattress! The warm sun pours down on you, it is true, but the cool breeze plays about your face and tempers the ray; and as you gaze upward into the unfathomable sky, and feel its pure cloudless azure penetrate your soul, and inhale the aromatic odour of the opening buds and the mingled perfume of a thousand humble flowers around, you fancy, for the time at least, that no couch in the world could yield you so refreshing or so delightful a repose.

Hereabouts we obtained a view of the beach far below, covered with huge rounded boulders of granite, all invested with a coating of green seaweed; for the tide was now at its lowest. The eye, roaming over the intermediate slope of fern, so feebly appreciated the distance, that it seemed an easy matter to run to its edge, and then scramble down the face of the perpendicular cliff, which appeared only a few yards high. The boulders upon the beach, too, appeared not too large or weighty to be turned over by hand, and I was actually meditating an attempt to explore the inviting locality, in hopes of finding many Annelides and Crustacea under those stones. But our more experienced friend assured us that those green-clad boulders were masses of many tons' weight; that the cliffs were from fifty to a hundred feet high, and so inaccessible that it would be utterly impossible to ascend or descend them unassisted. "Not long ago," said he, "a vessel came on shore in that very spot: walking here one morning early I discovered her on the rocks; she was a Norwegian brig in ballast, outward-bound; all hands were saved, but it was only by means of ropes passed down to them by our people, by which they were hauled up those cliffs that you think so easy to climb."

We now came to the Half-way Wall, so called because it cuts the island transversely in the middle. Its eastern extremity, close to which we stood, terminates in a huge mass of granite, on which a cubical (or rather parallel-sided) block, about fifteen feet high by eight wide, stands. It was formerly a true loganstone, being so poised by nature that it could be rocked by the hands of those who had nerve enough to stand on its narrow and lofty base, as our friend had often done. Now, however, it has slipped out of its equilibrium into a crevice, and is immovable; the action of the weather, as is supposed, having worn away its base.

The paths through the heath, and the open spots in many places, showed the power of atmospheric action to change the condition of the solid rock. These were covered with a sort of gravel, composed of white fragments about the size of peas, very uniform in appearance, which, when examined, proved to be nodules of quartz, liberated by the natural disintegration of the granite. A large quantity might be collected with little expense of time or labour.

An attempt was made to use the granules as gravel for garden walks, for which their regular size and form, and their pure white colour, would have made them very suitable; but the absolute want of any adhesive principle caused them to be rejected on trial: in technical phrase " they would not *bind*."

LUNDY ISLAND—(*continued.*)

THE TEMPLAR ROCK.

A LITTLE beyond the Half-way Wall we were introduced to "The Templar," a colossal human face in
profile, sculptured by nature out of the rock. It
forms a projecting point, one of those corners which,
from the southern end of the island, we see standing
out at the upper extremity of the bevelled slope; an
enormous block of granite, rudely split and shivered
by the elements, but accidentally fashioned, as you

look at it in bold relief against the sky, into so perfect a resemblance to the features of a man, that one can scarcely believe that it has not been touched by an artist's chisel. The features are bold and masculine, the nose sharply aquiline, the mouth compressed with a determined expression, the forehead projecting, the chin a little double, the neck muscular and swelling; the head is covered with a low round skullcap, furnished with a projecting peak in front: it requires, indeed, no stretch of fancy to imagine we see in it the portrait of one of those warlike Knights of the Temple, to whom the island at one period belonged.

We noticed here a curious phenomenon, with which our prolonged stay on the island made us sufficiently familiar afterwards. On looking back to the southward, we perceived everything distinct and palpable, except the lighthouse, the summit of which was enveloped in a semi-transparent haze, that streamed off some distance to leeward like a white veil. We were informed that it is a common thing for the fog to lie on the heights of the island, while the sides, the beach, and the sea, are perfectly free from cloud: hence the elevated parts are generally moist; and thus, doubtless, those springs are fed which issue from these lofty moors and trickle down on either side.

Farther on, still keeping along the inner or eastern side, a romantic glen opens, very boggy, and therefore difficult to be explored. By means of the tufts of a

sort of grass that grows here, however, we managed
to make our way some distance down it.   This plant
grows in large stools or tussocks, formed of the densely-
matted leaf-bases of successive seasons ; some of which
are eight or ten feet high, and two feet in diameter.
An agile person might leap from one of these to
another, and so traverse the valley without wetting
his feet.   Through the gully we had a view of Gannet
Cove, as also of Gannet Rock, an insular mass lying
off one of its points ; and here we saw the first outpost of the grand army of birds that we had come to
visit.   We pushed up on the opposite side of the
valley, through the tall fern, which was growing excessively rank, reaching about as high as our heads ;
sat down a few moments to rest, and amused ourselves by seeing which could cut the fern-stalk so as
to produce the most effective royal oak.   Perhaps
some of my readers may like to amuse themselves in
the same manner: if you have never seen it done,
select a stout leaf of the common brake-fern and pull
it up from the roots, then with a sharp knife cut the
stalk across slantwise, in the black part that is ordinarily immersed in the soil, when the section of the
vessels will display a very pretty semblance of a well-
grown oak-tree, either tall or widely-spreading, according to the direction in which you make the cut.

In the vicinity we found some interesting plants.
The beautiful blue skull-cap was growing in the
streamlet that trickled into the gully: higher up the
pretty little yellow pimpernel, or wood loosestrife, was

abundant; and so was the bog pimpernel, as, indeed, we found it widely spread over the north end and centre of the island. Mr Heaven mentioned his having met with the much rarer blue pimpernel on some former occasions; but it did not occur to us. Among the brake the wild hyacinth yet lingered in flower, but was found more numerously in fruit. The dwarf red-rattle, a lowly denizen of waste places, scarcely rearing its rosy vaulted head above the level of the moss in which it grows, occurred here, together with its usual companion, the bird's-foot lotus. The small upright St John's wort, an exquisite flower, the tiny eyebright, and the milkwort of the rich blue variety, were also among the plants we gathered here.

But now we were approaching the scene which had been the chief object of our curiosity. Near the northern extremity of the island stands a huge oblong block, like a square column, called the Constable: we pass this, and the wondrous spectacle suddenly bursts upon us. Much as our expectations may have been excited, they were in nowise disappointed; though my companions were not like myself naturalists proper, we were unanimous in declaring that the sight was more than worth the voyage, sea-sickness and all; it was a scene, the witnessing of which must always stand out prominently in memory, as one of the remarkable things, of which an ordinary life can reckon but few.

We turn the corner of a pile of rocks, and we stand in the midst of myriads of birds. We are on an in-

clined plane, extending, perhaps, half-a-mile down to
the sea-cliffs, composed of numberless hillocks of red
earth, on which lie, heaped irregularly, and partially
imbedded in the soil, great boulders of the granite
rock. On these, on the hillocks, and in the hollows
between, sit the birds, indifferent to our presence,
until within two or three yards of them, when they
turn the large liquid eye towards us, as if demanding
the meaning of the unwonted intrusion. If we avoid
sudden motion, we may approach still closer; but
generally at about this degree of proximity the little
group congregated on the particular stone or hillock
leap up, spread their short feeble wings, and fly with
a rapid laborious beating of the air, out to seaward.
The flight is painfully feeble at first, but presently
gathers strength and becomes more forcible, though
always fluttering.

The great congregation of birds begins just hereabouts; the cover of fern to the southward, which we
have been skirting, is not suitable to their habits; but
it extends forward as far as the eye can reach, and
is not then bounded, but spreads on around the north
extremity of the island, far down on the western side.

The air, too, is filled with them like a cloud.
Thousands and ten thousands are flying round in a
vast circle or orbit, the breadth of which reaches from
about where we stand to half-a-mile seaward. They
reminded me strongly, with their little wings stretched
at right angles to their bodies, painted in black against
the sky, of the representations we see in old astronomi-

cal works of the fixed stars arranged in the Ptolemaic
system in a crowded circle around the sun and planets.
If you attend only to those near you, they seem to
rush on in one direction in an unceasing stream; and
you wonder what can be the purpose, and what the
terminus, of the universal migration; but when your
eye has followed them a little, you perceive the circular movement, that the same birds pass before you
again and again, as they come round in their turn,
like the movers in a theatrical procession, that cross
the stage and pass round behind the scenes to swell
the array again.

But the earth and the air are not the only spheres
occupied by these birds: look down on the sea; its
shining face is strewn, as far as you can discern anything, with minute black specks, associated in flocks
or groups, some comprising few, others countless individuals. These, too, are the birds, busily employed
in fishing for the supply of their mates and young, or
resting calmly on the swelling undulation.

The fearlessness manifested by those that are sitting
around us, permits us to observe them at leisure.
They are principally of two kinds; the smaller has a
round large head, with a beak monstrously deep and
high, but thin and knife-like; and as if to make this
organ more conspicuous, it is painted with red, blue,
and yellow. The legs and large webbed feet are
orange-coloured; and these, too, are sufficiently
remarkable in flight, for the bird stretches them out
behind, somewhat expanded at the same time, in such

a manner that they appear to support the short tail,
the broad feet sticking out behind.   The whole of the
upper plumage is black; the face, sides of the head,
and under parts, pure white, except that a black collar
passes round the throat.   These are known by the
fishermen as sea parrots or coulternebs; but are more
generally designated in books as puffins.

The other species is larger, being nearly as big in
the body as a duck, but shorter in the neck.   The
beak is formed on the same model, but is more
lengthened; and it, as well as the feet, is black.   The
general proportions are more those of ordinary birds;
and though the distribution of the hues of the plumage, black and white, is nearly the same as in the
former species, the black, covering the whole head and
neck, combines, with the other differences I have mentioned, to render the discrimination of one from the
other easy, even at a great distance.   This is the
razor-billed auk.

These two species furnish the majority of the individual birds that are congregated just here.   But
when we get round yonder point we shall open the
haunts of several other kinds, almost as innumerable
as these.   It must not, however, be supposed that
they keep their localities so strictly as not to inter-
mingle in any degree.   From the point where we stand,
we may with a little care be able to discern individuals of all the kinds, more or less numerous.   The
different species of gulls, in particular, amounting to
four or five, are conspicuous for their long pointed

wings and elegant sailing flight. They are wary and
alert; we do not see them sitting still as we approach,
as the puffins and razor-bills do, for before we can get
within gun-range they are on the wing. Then, as
conscious of their powers, they are bold; sweeping by
over our heads, with a querulous scream; now and
then swooping down and making as if they would dash
at our faces, but taking care to swerve as they come
close, and gliding away with the most graceful ease
and freedom.

Let us examine for a moment the ground beneath
our feet. We need caution in moving about, for the
tussocks and mounds feel precariously hollow and
spongy; now and then the foot breaks through, and
the whole leg is buried in a dusty cavity that gives
forth an insufferable odour of guano; then as we
jump on a hillock, it totters and breaks off from its
base to roll down the hill, laying bare an interior
riddled with holes like a honeycomb. These hillocks
themselves are nothing but enormous tufts of the
common thrift or sea-lavender, so often used for
edgings in cottage gardens: the plant in a succession
of years assumes a dense hemispherical form, while
the decay of the old leaves forms a reddish spongy
earth, which constantly accumulates, and constitutes
the soil on which the living plant grows.

Under the projecting shelter of one of these tussocks
we found a nest of one of the gulls, the lesser black-
backed species as was supposed. It was a platform
made of the red leaf-bases of the thrift, dry and brittle,

on which lay one young chick and one egg. The
latter was larger than a hen's egg, of a dark greenish
hue with black spots; it was on the point of hatching,
for I distinctly heard the feeble piping of the impa-
tient chick within, whose beak had already begun to
chip the shell. The hatched young one, a tiny creature, covered with pale-brown down, lay quite still
with shut eyes, which it opened for a moment when
touched, to close them again in stoical indifference.

Presently we came upon another nest, containing
one young rather more advanced; its clothing of down
prettily spotted with dark-brown. Then another
with two eggs of a dirty white, mottled and splashed
with brown, which was conjectured to belong to the
glaucous gull, a powerful and handsome bird seen
hovering about, of snowy-white plumage, except the
back and wings, which are of a delicately-pale bluish-
gray.

The whole atmosphere was redolent with the strong
pungent odour of guano, which, as everybody knows,
is the excrement of fish-eating birds, collected from
the rocks on which they breed, where it has accumulated for ages. The same substance was splashed
upon the stones and earth wherever we looked; we
saw it falling through the air; our clothes were spotted
as if with whitewash; and we scarcely dared to gaze
upwards on the circling flocks, lest our eyes should
suffer the misfortune of Tobit.

It is to the puffins that the burrows with which the
soft vegetable earth is honeycombed are chiefly attri-

butable. The whole island is indeed stocked with rabbits, and their warrens (or " buries," * as the local phrase is) are very numerous. The puffin does not hesitate to appropriate these whenever he can ; but as there are many more birds than beasts, the former are generally compelled to excavate for themselves : this is effected by means of the powerful cutting beak, to the depth of two or three feet. At the bottom of the hole the egg is laid, never more than one. We saw several egg-shells, from which the young had been hatched ; they were nearly as large as hen's eggs, of a dirty whited-brown tint, which is said to be derived from the soil, as they are purely white when first laid. We had no means of digging them out, and we did not choose to explore the burrows by thrusting in our arms ; for the puffin, if at home, would have given our intrusive fingers such a welcome with his strong and sharp beak, as we might not soon have forgotten.

Mr Heaven informed us of a curious habit in the economy of these birds. Immense numbers come to the breeding-place in April, to reconnoitre the ground : they remain three or four days, then disappear so completely that not a single bird is to be seen. In about a fortnight they return for good, and set about the work of family-rearing. Then mortal combats may be witnessed ; the rabbit and the puffin fight for possession ; the old buck stands up in front of his hole,

* Doubtless for "burrows ;" but the word is pronounced as we ordinarily pronounce the verb "to bury" (the dead). Probably we have but two forms of the same word in "bury" and "burrow."

C

and strikes manfully, while the knife-beak of the dishonest bird gives him a terribly unfair advantage. Sometimes two male puffins contend; each strives to catch his adversary by the neck; and when he can accomplish it, shakes and holds him with the tenacity of a bull-dog.

Auks and guillemots likewise bear a part in the exploratory April visit; but not in such numbers as the puffins.

One of our party knocked-over a puffin with a clod of earth, just to examine it. We did not wish to destroy them, and therefore abstained generally from throwing. It was stunned, and lay in our hands while we admired the thickness and closeness of its plumage, beautifully clean and satiny, especially the white parts. Presently it began to open its large dreamy grey eyes, so singularly set in scarlet eyelids : we did not wish to prove the keenness of its beak, and therefore laid it on a rock in the sun, where no doubt it soon recovered.

It must not be supposed that this was any feat of skill in the marksman. It would have been perfectly easy to procure hundreds in the same way. Our friend assured us that he had himself knocked down six with one stone; and that he had seen twenty-seven bagged from a single shot with an ordinary fowling-piece, not reckoning many more which were knocked over, partially wounded, but which managed to fly out to sea.

We walked on a few rods farther. The character of the declivity continued pretty much the same; but

we had opened a point of the distant cliff which was
cut into a series of rocky ledges, like a wide flight of
steps leading to some magnificent building. On these
were seated a dozen or twenty gannets, beautifully
snow-white birds, with black tips to their wings, larger
than geese. We could easily have scrambled to their
rock, but our friend was reluctant to have them disturbed. This fine bird used to be numerous here;
and Gannet Cove and Gannet Rock received their
appellations from the hosts of these birds that used to
make that neighbourhood their resort; but having
been much annoyed by idle gunners from the main,
they had deserted the island, it was feared finally.
Lately, however, a few pairs have returned, to the
gratification of the proprietor, who is desirous of their
increase. In truth, they are noble and beautiful birds;
their long pointed pinions enable them to wheel and
glide about in the air, to soar aloft, or swoop, or float
on motionless wing at pleasure with the utmost grace;
while the contrast of the black wing-tips with the
general whiteness of the plumage, cannot fail to elicit
admiration. As they sail near, we perceive that the
neck and poll are tinged with buff; but this exception to the general purity of the vesture is not at all
conspicuous, nor universal. Their cry is, " crak, crak,
crak," uttered on the wing. The snowy purity of the
mature plumage is said to be reached through several
alternations of opposite hues. The young, when newly
hatched, are black and quite naked: their first coat of
down is white; this is succeeded by a plumage of

black slightly spotted with white; and this by the spotless white investiture in which we saw them.

Another reason why the gannets should not be disturbed, while so few as they yet are, is the bold piratical character of the larger gulls. These are ever on the watch to destroy the eggs of the gannet, the moment both the parents are flown. We had a proof of the ferocity of these predaceous birds before our eyes. As we were looking down the slope, we saw a glaucous gull emerge from a puffin's hole into which he had just crept, bringing out the little black puffin-chick. We watched the marauder. shake his victim and give it repeated blows with his beak, the poor little thing now and then crawling away feebly, just as a mouseling does when half killed by a cat. We began to run towards the spot, the gull taking no notice till we got pretty near, when he turned up his eye and gave us a look of impudent defiance, then deliberately seized his prey in his beak, and bore it off triumphantly far out to sea. The larger gulls will sometimes swoop down upon a group of puffins sitting on the sea, and snatch up an adult from the flock in the powerful beak. Mr Heaven has seen this done.

Our attention was here pointed to a new bird. On the lower ledges of the wide stair-like rock occupied by the gannets, sat, in little crowded rows, many birds about as large as pigeons, which in form and in the colours of their plumage they much resembled. They were the kittiwake, the smallest of the gulls that can properly be called indigenous to our shores. We afterwards made closer acquaintance with the species.

The shearwater is said to breed in the rocks hereabouts; but we did not notice it, nor do I know of which species it is. Nearly at the edge of the slope we observed a stout iron rod erected, standing some ten or fifteen feet high. On inquiry we found that this, with a corresponding one at some distance, is used for the support of a long but narrow net, which is stretched along like a wall at the edge of the precipice, to intercept the puffins. These birds, when they fly, shoot down in a straight line, just sufficiently above the ground to clear the rocks and hillocks; they thus strike the net, and are caught. They are also taken in numbers by dogs, which run upon them before they have time to fly; and in other modes, chiefly for the sake of their soft and abundant feathers.

From the spot where we now stood there extended a considerable space, almost covered with the wild hyacinth, as we could see by the fruit-bearing stalks. The contrast which this large belt presented when in flower, with the thrift which occupies as exclusively the range below it, was described to us as very curious and pretty; the whole forming two parallel zones, the one of blue, the other of pink. Large beds of coarse sorrel were prominent in the vegetation here; and the crevices and bases of the rocks were fringed with the singularly-cut leaves of the buck's-horn plantain, growing in unusual luxuriance. The pungent, peppery scurvy grass we also found very fine.

We now approached the north-west point, the very
extremity of the island; no slope of earth, but a wilderness of huge castellated masses of granite, piled
one on another in magnificent confusion.   By scrambling between and over these, we contrived to take a
perch, like so many of the tenant-birds themselves, on
the very verge of the stony point, whence we could
look over on each side, and gaze on the boiling sea at
the foot of perpendicular precipices.   In truth this
was a noble sight; the point was fringed with great
insular rocks, bristling up amidst the sea, of various
sizes, and irregular angular shapes, partially or wholly
covered by the tide at high water, though now largely
exposed.   There was a heavy swell from the westward, which, coming on in broadly heaving undulations, gave the idea of power indeed, but of power in
repose; as when the lion couches in his lair with
sheathed talons, and smoothed mane, and half-closed
eyes.   But no sooner does each broad swell, dark and
polished, come into contact with these walls and towers
of solid rock, than its aspect is instantly changed.
It rears itself in fury, dashes with hoarse roar, and
with apparently resistless might, against the oppo-
sition, breaks in a cloud of snowy foam, which hides
the rocky eminences, and makes us for a moment
think the sea has conquered.   But the next,—the
baffled assailant is recoiling in a hundred cascades,
or writhing and grovelling in swirls around the feet
of those strong pillars, which still stand in their majesty, unmoved, unmovable, ready to receive and to

repel the successive assaults of wave after wave, with ever the same result.

We watched the war of the elements, the conflict of Land and Sea, a while, with somewhat of the interest that attaches to a doubtful combat, though we well knew the fortress could not be taken by assault; and at length we turned to other features of interest which our vantage-ground commanded.

Looking over the battlemented margin of the platform on which we stood, we could see the entrance of a fine cavern, sixty feet in height, about thirty in width, and perhaps eight hundred in length. It completely perforates a projecting promontory, the part of the coast, indeed, which we had been skirting, on which our principal observations on the birds had been made. A boat can go right through, but only at high water, because there is a rock in the midst of the course, which, at any other state of the tide, leaves too narrow a channel on either side. But the most interesting fact connected with the cavern is, that a spring of fresh-water is said to rise in its centre, bubbling up through the sea-water that overlays its mouth. Mr Heaven could not vouch for this on personal observation; but the well-known occurrence of similar phenomena renders credence in this case no great difficulty. The breaking of the sea into the mouth of the cave, narrowed as it is, and the reverberation of its hollow roar from the sides of the chasm, were particularly grand and striking.

When our first emotions of admiration at the grander

features of the scene were a little exhausted, we had
leisure to look at the living occupants of the rocks.
The perpendicular cliffs of the naked rock, broken
into vast angular masses, square columns, and buttresses, like the walls of some old irregular castle, and
cut into shelves and ledges, sometimes only a few
inches wide, presented a very different scene from the
sloping wilderness of thrift-tussocks, interspersed with
boulders, which we had seen tenanted by the puffins
and razor-bills. Both of these species, indeed, were
found here also in considerable numbers; but the
species more strictly appropriated to this locality was
the foolish guillemot, or " mer," as it is better known
to the fishermen. All along the little ledges, around,
above, and beneath us, we saw the guillemots sitting
in rows, row above row, almost as close as they could
comfortably place themselves, every one bolt upright,
the manner of sitting common to the puffins and
razor-bills also, but not to the gulls or gannets, which
incline the body when resting, as most birds do. It
is the position of the feet, set far behind, in the short-
winged plunging birds of the diver and auk families,
that makes the upright posture that of rest, this being
the only manner in which the centre of gravity can
be brought over the feet. The whole sole rests on the
ground, and not the toes only, as in other birds.

Many of these birds were incubating, and others had
a chick. Not the least vestige of a nest was there, not
a fragment of seaweed, not a leaf of thrift; the single
egg, never more, is dropped on the smooth shelf of

stone, perhaps not wider than its own length, where one would suppose the first puff of wind would roll them over the edge, and involve them by scores in the irremediable fate of Umpti-dumpty in the nursery-rhyme. Still we did not discern on the groins and points of the rock below any spatterings which would indicate the frequency of such an accident; nor can we suppose, from what we know of the economy of the works of God, and of the almost infallibility of instinct, that it is at all common. Probably the egg is rarely or never left unprotected, except in unwonted circumstances, one parent relieving the other in incubation; and we could see how cleverly the old bird kept its frail charge between its legs, even as it moved to and fro. An intelligent observer of animals, who is very familiar with these birds, told me that he had seen a gull attack a sitting mer with the design of robbing her of her egg. They engaged stoutly, the mer pushing her egg behind her, while she faced her enemy. At length she caught him by the leg, and pinched so hard, and held on so firmly, still all the while covering her egg in the angle of the ledge, that at length she fairly drove the robber off.

The chick does not sit between the feet of the parent, but cowers beneath one of its wings, which is drooped to shelter it,—a touching sight, as every manifestation of parental care and affection in the inferior animals is. If the account which the fishermen at Flamborough Head gave Mr Waterton is correct, and there is every reason to credit it, the young

are indebted for their first introduction to the sea to the
parental care displayed in a very interesting manner.
They take to the water and fish for themselves long
before they are able to fly; and as they would inevitably be killed on the sharp points of rock if they
attempted to fall or leap down, the parent invites its
offspring to climb on its broad back, and thus carries
it down.   This we did not see; but we were witnesses,
in plenty of instances, of the prompt and ample supply of food brought by the industry of the parent bird,
either to its sitting mate or to its unfledged young.
The air here, as on the other side, was filled with
birds on the wing; and the sea below, not amidst the
boiling eddies of the rocks, but outside, was even
more densely crowded with swimmers; and ever and
anon one would shoot by us with several little bands
of silver depending from its beak, the fruit of its suc-
cessful efforts.   These are invariably carried, no matter how many they may be, transversely, held fast by
the head, the body hanging down.   When we remembered that each fish must be caught separately,
we were at a loss to understand how the first captured
could be retained in the beak in this orderly manner,
or, indeed, how held at all, while another was seized.
Would not the first fall in the act of opening the
mandibles a second time?   One of the party, with
his fowling-piece, brought down a guillemot, returning with prey, and an examination appeared to me to
resolve the difficulty.   Ten little sandlaunce this ill-
fated mother was bringing to her chick, when the

leaden shower overtook her. On opening the mouth
I perceived the tongue large and muscular, and its
edges cut near the base into sharp teeth, pointing
backward. I have no doubt that each fish, as taken,
is placed between the tongue and the upper mandible,
and firmly held by these serratures, while the lower
mandible is allowed to open freely for the seizure of
another, which, in turn, is secured in the same manner, until a sufficient booty is collected to fly home
with.

The young of the sandlaunce, and a small fish
called "brit," which Mr Heaven believes to be the
fry of some species of the herring family, form the
favourite prey of all these birds; and the rough water
off this north-west point is the favourite fishing-ground
for them. A very strong tide runs round this end of
the island, the strongest in the whole channel; hence
a "race" is almost always running; that is, a violent
agitation of the water, a strong ripple in calm and
smooth weather, and what seamen call a "bobbery,"
a tossing, breaking sea, when there is anything of a
swell on. The fish-fry delight in such a race, and
are pretty sure to be found there in shoals.

The egg of the guillemot is large for the size of the
bird, and of so unusual a form, that when once seen it
is never likely to be mistaken for any other. It is a long
cone, with both ends rounded. Its appearance is striking and *bizarre;* the ordinary ground colour being a
fine green, variously splashed and spotted with darker
green or black. There is, however, much diversity in

the colour both of the ground and of the markings;
and, indeed, in the shape, though the characteristic
form is generally maintained.   The eggs are taken in
considerable numbers by the youths on the island, as
well as by fishermen from the neighbouring coast.
The explorer and collector is let down from above by
a rope in the hands of his comrades; and, as he tra-
verses the ledges, he picks up the eggs, and places them
in a large pocket tied round his waist.   In the season
we see them offered for sale by the fishermen's children at Ilfracombe, at a penny each; and many are
purchased as curiosities by visitors, who are struck
with their singularity and beauty.   If I mistake not,
I have seen them sold also in the streets of London
by sailors.   In Newfoundland I have often eaten
them, where they are well known by the name of
Baccalao-birds' eggs.   Their taste and flavour are by
no means unpleasant; but the glair, which remains
semi-transparent, has a curious appearance.   On Lundy
they are used in the preparations of cookery, but are
eaten alone only by the poor.

That rarest of British birds, the great auk, a species as large as a goose, there is some reason to believe,
is occasionally seen at Lundy.   A specimen was picked
up dead in the sea near the island in 1829; and the
fishermen have spoken to Mr Heaven of having seen
at the herring-station an auk of very large size, which
that gentleman has conjectured to be the species in
question.

LUNDY ISLAND—(*continued.*)

THE MOUSE-TRAP AND MOUSE-HOLE.

Two curiosities were proposed to us to be visited on the third day. The one was called the Devil's Lime-kiln, the other was the Seal Cavern.

The morning rose in that cool and cloudless brilliancy which so often characterises the opening day at this lovely season. On the preceding evening, one of

us, looking on the gorgeous western sky, had hope-
fully said, in the words of Shakespeare,—

> " The weary sun hath made a golden set,
> And, by the bright track of his glitt'ring car,
> Gives token of a goodly day to-morrow."

And now the morrow was come, and the promise
was not broken. Hope and hilarity were strong in
each of our minds, as we rapidly completed our slight
preparations for the morning's jaunt, and awaited the
arrival of our kind guide; and I fear none of us were
able to sympathise very deeply with the sorrow of the
old farmer, who was bemoaning the loss of a thriving
young bullock, that had just been found dashed to
pieces at the bottom of one of the frightful precipices
that form the north-west edge of the island. These
casualties, however, are reckoned among things regular
and to be expected in Lundy husbandry. Some two
or three of the young cattle and horses are lost every
year from this cause. They incautiously feed close to
the edge, when a puff of wind catches them on the
broadside, and over they go; to the no small joy of
the carrion-crows, who flock to the funeral feast.

At length, away we sallied through a gate at the
rear of the farm, across wide, moory fields, till we
struck a broad road, marked off by stone posts at
regular intervals, each bearing conspicuously painted
the letters T. H. Our curiosity was excited by the
boundary-stones ; and we were informed that the
ground so marked off is the property of the Trinity
House, forming a road thirty feet wide, and about a

mile in length, leading from the beach where we
landed to the lighthouse. This road, and the ground
on which the lighthouse stands, form the only exception to the sovereignty of the island.

These boggy, elevated moors presented us with the
yellow blossoms of the great spearwort (*Ranunculus
lingua*); the rough water-bedstraw (*Galium Wither-
ingii*), with its whorls of curious leaves, beset all
round their margins and along the backs of their
nervures, as well as the edges of the angular stem,
with minute, barbed prickles, that catch the finger as
it is passed up the plant, was likewise abundant here.
The bog-pimpernel (*Anagallis tenella*), a lowly but
lovely little plant, was likewise profusely strewn over
the spongy moors, its sweet little pink blossoms occurring at every step.

Close to the south-west corner of the island we
came rather suddenly upon the first object of our
curiosity. In the midst of the heath-covered slope
yawned a terrific chasm, into which it made us
shudder to look. Its form is irregularly square at
the top, where it is about two hundred and fifty feet
wide. The sides in some parts are quite perpendicular,
but gradually approach each other to the bottom, so
as to resemble a funnel, which we judged to be about
as deep as the mouth is wide, or about two hundred
and fifty feet. The edges and sides of this fearful pit
are fringed with a scanty but various herbage, among
which we noticed many plants in flower. The upper
parts were gay with the blue sheepsbit, and the flesh-

coloured stonecrop; the thrift, the bladder-campion, and the samphire, were springing out of the crevices, and the yellow blossoms of the long-rooted cat's-ear, closely resembling those of the dandelion, were mingled with them. On some of the ledges far down were growing large tufts of a coarse plant, which our friend informed us is occasionally used as a substitute for spinach; we could not get near enough to examine it accurately, but it was probably one of the goosefoots.

The distant bottom of this hole was strewn with large blocks of alabaster, some of them twenty feet high. Among these there is, at one side, a narrow, door-like opening, which leads, by a natural tunnel, to the beach at the foot of the cliffs. This affords the only means of access into the chasm; and is, from the precipitous character of the coast, available only with a boat, and in calm weather, for when there is any swell the sea dashes furiously into the tunnel.

One part of the margin of the chasm forms a slender ridge like a wall, dividing it from a very steep declivity; along this precarious path one or two of our party scrambled on hands and knees, to gain a better view of the recesses of the abyss. While we were thus engaged a falcon flew out, whose red back and wings, as he emerged into the sunlight, shewed him to be the kestrel: he hovered a while in the air over his den, facing the wind like a ship at anchor, in that peculiar manner which has obtained for this bird the appellation of windhover.

We turned our gaze seaward. There we beheld a

vast cone of granite, almost insulated from the shore.
The fishermen and inhabitants believe that this rock,
if it could be turned over into the limekiln, would
exactly fit and close it. Hence they have named it
the Shutter Rock.

The comparison of this deep pit with an orifice at
the bottom to a limekiln is striking and felicitous;
but why it should bear the devil's name I cannot
understand. The habit, which prevails in all parts of
the country, of associating the great adversary of God
and man with those phenomena of nature which are
vast, or grand, or terrific, is both preposterous and
repulsive. It originated probably in the darkness of
the Middle Ages, when mankind were ready to attri-
bute to Satan operations with which he had nought
to do, yet strangely forgot his power as the great
tempter to sin, and overlooked the real work in which
he is ever engaged, of " blinding the minds of them
which believe not, lest the light of the glorious gospel
of Christ, who is the image of God, should shine unto
them" (2 Cor. iv. 4).

We threw ourselves down on the purple heath and
the soft beds of wild thyme, that covered the broad
slope between the limekiln and the edge of the cliffs.
The sun was pouring down his fervid rays upon us as
we reclined, and his disk of brightness was reflected
in thousands of rippling waves, from the wide expanse
of sea that lay stretched between us and the undulating line of blue coast opposite. Just over against
us, some five leagues distant, was the promontory of

D

Hartland, with the picturesque little watering-place
of Clovelly; from whence headland after headland, on
the one hand those of Devonshire, on the other those
of Cornwall, receded into a dim and undistinguishable haze.

Insect life was active and busy around us. Little
beetles, whose coats sparkled in the sun, were crawling on the herbage; a tiny attelabus, of coppery
lustre, seemed rather common; the lovely green cicin-
dela, sometimes popularly called tiger-beetle, from its
beauty and voracity, was seen, but was much too agile
and wary to be caught; and the rose-chafer, that
peculiar accompaniment of a summer's noon, was
buzzing like a bee among the flowers. Butterflies,
too, of various species, were flitting to and fro; the
large and small garden-whites were perhaps attendants upon man, as the cultivator of pot-herbs, their
proper food; but others were indubitably indigenous.
The meadow-brown and the little gate-keeper were
pursuing that low, dancing, jerking manner of flight,
close to the turf, that distinguishes the genus to which
they belong; the tiny alexis was opening and shutting
its azure wings in the sun, as it sat upon the flowers,
as if inviting capture, but darted away when approached, with a swift, wheeling flight, and playfully
returned to the same flower again. And we saw
a rarer insect than any of these, the painted lady,
come fluttering by on vigorous wing, and shoot away
like a meteor.

Gulls were screaming in the air around, and cir-

cling about the cliffs; troops of guillemots were perched
upon the ledges, one and another every instant dropping down, like an arrow, into the sea, and presently
returning with the captured prey; and upon the sharp
edge of an insular rock outside the Shutter, known
as the Black Rock, sat a row of cormorants, preening
their glossy plumage after the morning's meal.

We rose and pursued our sinuous way along the
turf, by the margin of the precipitous cliffs of granite.
A little to the north of the Limekiln we came suddenly on the edge of a deep cove, at the mouth of
which rose an enormous mass of rock, with walls as
steep as those of a church, called Goat Island. It
was the scene of a fatal accident not long ago. A
party had come over to visit the island as we had
done. A young man of their number must needs try,
in spite of warning and entreaty, to climb Goat Island, with no other purpose than that of displaying
his agility and his hardihood. He had proceeded
some distance up the dizzy height, when, his foot
slipping, he fell on the stones beneath, and broke his
back.

Into this cove we descended by means of the round
and soft, yet sufficiently firm, hillocks of thrift, jump-
ing from one to another. When these ceased, we had
to scramble down by the fissures of the rock, until we
came to a cyclopean wilderness of huge blocks and
boulders of granite, strewn over the bottom, and piled
one upon another, in grand confusion. They were
worn smooth by the action of the waves, which had

been beating on them perhaps for ages ; and the lowest of them were rendered still more slippery by the drapery of green and olive sea-weeds (*Ulvæ* and *Fuci*) with which they were covered. It was, therefore, unpleasant and difficult, not to say hazardous, to make way among them by climbing over the masses, creeping under and between them, and leaping from one to another.

Nor was there much, on such a shore as this, either of zoology or botany, to reward the search. Professor Harvey has truly observed, that " on a shore composed of granite rocks, where the masses are rounded and lumpy, with few interstices or cavities in which water will constantly lie, and presenting to the waves sloping ridges along which the water freely runs up and down, very few species of sea-weeds, and these only of the coarsest kinds, are commonly to be met with."*

However, we had had the pleasurable excitement of overcoming the difficulties of the descent and the exploration, and we had now to essay those of the ascent. When we arrived at the top, our clothes and hands were perfumed with the strong odour of the milfoil, through whole beds of which we had been penetrating ; we found ourselves, moreover, nearly wet through with the moisture which yet loaded the herbage, from the dense fog of the preceding night.

A mid-day dinner left the afternoon free for a visit to the Seal Cave. A council was called on the practica-

* Sea-side Book, p. 54.

bility of effecting an entrance, and on the best mode
of gaining access to it. Old Captain Jack and his
son, Captain Tom, agreed in thinking that the low
state of the water, for it was now spring-tide, would
permit our approach to the cavern on foot, but that
the surf would render it difficult for a boat to land,
which otherwise would have been the most pleasant
mode of reaching the spot. It was, therefore, resolved
that we should approach it from the landward side,
descending the cliffs at Benjamin's Chair. We wended our way, accordingly, as if we had been going to
the Castle, but turning short to the right, we found
ourselves at the edge of the precipice, in the middle
of the south end of the island above a shallow bay
called Rattle's Landing-place. A line, drawn from
this spot to the landing-place on the eastern side, divides the island geologically. All to the north of this
line, including the greater part of the island, is granite; the little corner to the south-east of it is the gray
friable shale, common to North Devon. The junction
of the two structures is well defined down the cliff.
At the point of union copper ore has been found, in
sufficient quantity to warrant the formation of a shaft,
the erections of which were pointed out to us.

A narrow track, easily overlooked by those who are
not familiar with it, leads down to a little grassy platform. A huge perpendicular wall of granite forms
the back, thirty feet high, profusely clothed with gray
and orange-coloured lichens in loose shaggy tufts. A
semicircular horizon, dividing the blue expanse of sky

from that of the sea more deeply blue, was in front.
A magnificent scene it was in its grand simplicity;
nor unappreciated, for it was evidently a favourite
resort. A long tea-table, rudely made of unpainted
boards, which the sun had warped out of all shape,
had been set up under the rocks, and a bench on each
side afforded accommodation for a rather numerous
party. Nature had herself provided a throne of massive state, suited to the giant, whom imagination
might picture as the presiding genius of the place.
A square cavity in the granite wall formed a low-
seated chair, furnished with projections resembling
elbows, and a rest for the feet. This seat, which for
some reason or other, unpreserved by tradition, is
called Benjamin's Chair, gives name to the place.

While we rested here, Captain Jack appeared, followed by two servants bearing a long ladder, a lantern,
and a few tallow candles. We watched the proceed-
ings with interest. The assistants, having fastened a
long line to the ladder, go down with their charge;
the one letting it gradually down from above, the
other guiding it in its descent. Then down goes the
Captain with the lantern, and we all follow as best
we may; each one concentrating all his thoughts on
securing his own footsteps on the giddy height; for
we had to make a descent of four hundred feet, down
a cliff which, though not actually a precipice, was
fearfully steep. But we all contrived to scramble
down without injury, except a sting on the finger,
inflicted by a bee that considered himself insulted,

when one of our party thought to obtain a little assistance by grasping a tuft of thyme which the busy
insect had appropriated. " Take your time," said the
Captain. " I have not gained much by taking thyme,"
grumbled E., holding up his smarting finger.

A more efficient help was afforded by the angular
projections of the solid rock, which occurred here and
there, and, in one portion of the descent, by the sides
of a watercourse, which, though the roughness of the
way was increased by the rolled masses lying loosely
in it, was less perilous than the open declivity.

Sad witnesses to the power of the winds and waves
were lying in our way; for we saw, at a considerable
height above the bottom, the blocks and ironwork of
some ill-fated vessel, so firmly jammed into the crevices of the rock as to resist all efforts to dislodge them,
without more labour than they were worth. These,
as the Captain told us, were the relics of a fishing-
smack that was driven on the rocks below, of whose
hapless crew not one survived to tell the story.

Behold us then collected at the bottom, or as near
to it as we were destined to go; for, though it was
spring-tide, and the hour of low-water, no beach
appeared, but the clear transparent sea was washing
the foot of the cliff. On a narrow slanting ledge,
some eight or ten feet above the water-line, we were
all perched in a row, like so many guillemots; and
there we had quietly to remain till some needful pre-
liminaries were adjusted. We now perceived the use
of the ladder, which was not at all intended, as some

of us had naïvely supposed, to help us down the
declivity. The ledge on which we stood was not horizontal, but would have led us into the sea if we had
pursued it. At a certain convenient spot, therefore,
the ladder was set, and held firmly by the two men,
while we, one by one, *shinned* up to a higher ledge.
Along this we crept in the same manner, our feet
shuffling along the narrow shelf, our fingers hooked
into the crevices above; for these ledges were often
barely wide enough for the foot to rest on lengthwise.
As they all had a similar inclination, the same process
had to be repeated several times, the ladder enabling
us to mount to another ledge, when the one on which
we were walking dipped into the sea.

While holding on to the broad surface of the precipice, and especially in the moments occupied in waiting for the ascent of those who happened to be foremost in the line, it was interesting to look down
beneath our feet into the hollows between the rocky
masses, covered with water of crystalline clearness,
which rose and fell with every wave, but was prevented from breaking by the barrier of rocks outside,
on which the violence of the swell had spent itself.
In these hollows the large seaweeds were waving, the
wrinkled fronds of the oar-weed, floating like the
streamers of a ship, and the massive tangle tossing
about its long many-fingered hands, as if in distress,
with every undulation. .The submerged rocks, too,
were densely studded with the olive-coloured cups of
the sea-thong; many of which were crowned with the

singular appendages which bear the fructification; narrow forked straps or thongs, not more than a quarter of an inch in thickness, but stretching to a length of several yards, and springing from a point in the centre of each cup-like base.

After rounding in this manner the face of the cliff for a considerable distance, we came at length to some rocks which were high and dry above water, where, as we stood, the wide mouth of the dark cavern yawned immediately in front of us. Between us and it, however, lay an ample area, strewn with boulders of various shapes and sizes, but almost all covered with the sea, which was breaking over them with a formidable surge.

Now another council of war. How are we to pass this Scylla and Charybdis in one? The ladder comes again into requisition; when laid down horizontally, its extremities just reach across the space, from our position to a dry rock at the cave's mouth; its middle being supported by the top of a boulder which rose above the surface.

We looked rather blank at this precarious causeway; our only chance of getting over dry lay in the nimbleness of our heels; for every breaking sea washed away the ladder, despite the efforts of the servants to hold it firm at the ends. To him who was not agile enough to skip across in the interval between one sea and another, a ducking was inevitable.

By Captain Jack's advice, all of us took off our stockings and our upper garments, tucking up our trousers, and replacing boots and shoes, for the protection of our feet in crossing. Captain Jack remained on the rock, and became the depositary of clothes, watches, note-books, &c. "Here goes!" said one, and, rapidly stepping from rong to rong, adroitly effected the passage between the seas. "Oh, dear!" said another, "I can never do that." "I think *I* can," said a third; "I'll try at least." He essayed it, but was scarely half-way across when "Look out," was the cry; and a green curling wave at the same moment swept the ladder from the grasp of the assistants, and our luckless adventurer found himself, when the wave had passed over his head, up to the waist in water.

This was poor encouragement for the others, who, despairing of tripping it on such a light fantastic toe as the first had exhibited, determined to creep along on hands and knees, meekly resigning themselves to the brunt of the sea, with the philosophic exclamation, "'Tis only a wetting!"

When the hilarious mirth produced by these scrapes had subsided, we prepared to enter the cave. It was a noble vault, of sixty feet in height and twelve in width. For a little space we stepped over boulders; then a broad pool crossed our way, extending from wall to wall, seven or eight feet deep. Again the ladder was our medium of passage; now without risk, for the clear bluish-green water was unruffled as a mirror, and the narrow segments of the black tangle lay motionless in the depths, clothed with miniature forests of a tiny zoophyte, the delicate zigzagged *Laomedea*.

The damp walls of solid granite were studded with marine animals, but not nearly to the extent that I had anticipated. ' The low oval cones of the common limpet were adhering to the rock, with the little shelly tribes of *Serpulæ*, and small patches of orange and olive-coloured sponges ; and some parts of the sides and rocky floor were plastered over with what appeared a coating of brown mortar, but which, when examined, was seen to be an assemblage of tubular cells, composed of grains of sand, agglutinated together by an animal cement, so as to form walls of exquisite mosaic work. Each cell is inhabited by a worm (*Sabella alveolata*) of curious structure, and instincts no less remarkable.

After we had passed the pool, the bottom consisted of fine sand, wet but firm, its level sensibly rising. The cavern grew every moment darker and narrower, and here the candles were lighted and distributed. Each of us carried a piece in his fingers, which soon became streaked with stiffened streams of tallow, and one fragment was committed to the lantern as a reserve in case of accidents. Southey's fine description of such a cavern as this occurred to the mind :—

> " The entrance of the cave
> Darken'd the boat below.
> Around them from their nests
> The screaming sea-birds fled,
> Wondering at that strange shape,
> Yet unalarm'd at sight of living man,

Unknowing of his sway and power misused :
 The clamours of their young
 Echoed in shriller cries,
Which rung in wild discordance round the rock.
 And farther as they now advanced,
The dim reflection of the darken'd day
 Grew fainter, and the dash
Of the out-breakers deaden'd ; farther yet,
 And yet more faint the gleam ;
And there the waters, at their utmost bound,
Silently rippled on the rising rock."

      (" *Thalaba*," xii. 8.)

We proceeded silently and with caution, for we were now approaching the principal chamber, the place where seals would be found, if any happened to be at home. But in order to enter this hall we must pass through a gallery so narrow that a person could only squeeze himself along it sidewise. It is just as the foremost emerges from this passage that the seals make their rush. Alarmed by the approaching footsteps, they wait with expectant gaze until the intruder appears in their doorway. The sudden flash of light from the candle into their obscurity is the signal for their escape. With one bound the seal dashes at the man, who, if he be not thoroughly prepared for the shock, will inevitably be knocked over, while the seal makes good his exit across the prostrate person of his baffled invader.

All this was described to us while one of the servants, a cool, resolute fellow, used to the warfare, was exploring the passage, peering through the darkness, with his light above his head, and a stout bludgeon grasped in his right hand, ready for a blow. This man told us, as we returned, that he had killed no fewer than five seals on one occasion within the cavern.

We were not, however, favoured with so stirring a termination to our adventure. No sound proceeded from the interior, as our vanguard passed beyond our sphere of vision, and we all in succession followed him in.

We found ourselves in a gloomy chamber of spacious area, and so lofty that the united light of our feeble candles could not struggle to its roof. The walls were formed of the plain smooth rock, not particularly damp, and devoid of any incrustation or deposit of stalactite, the rock being composed entirely of granite, of which lime is no ingredient. There is a low and narrow hole at the farther end of the chamber, into which a man may enter by creeping on his hands and knees. It is believed to lead to another cavity, but none of us cared to explore it.

Our curiosity being satisfied, we commenced our return, which we effected in the same manner as our entrance, except that in crossing, by means of the ladder, from the cave to the rock where we had left the worthy old Captain, we were more unlucky, for every one was washed off from his hold by the surf. This involuntary bath, however, was no great misfortune; for the beams of the burning sun soon dried our drenched garments. Indeed, the contrast which we felt as we emerged from the chilly cavern into the

warm sunny air without, was like going into a bake-
house on a day in November.

I took the opportunity, before climbing the cliffs, of
examining the rock pools that were exposed by the
present low condition of the tide. It was evident
how much superior, as a field for the zoologist or
botanist, the shale is to the granite; for while the
latter presented no tide-pools, and comparatively few
of the finer or more delicate seaweeds, the former,
nearly clear of boulders, exhibited a comparatively
level surface, hollowed into numerous pools, varying
much in form, size, and depth. Though the aspect
was a southerly one, and much exposed to the sun's
rays, the seaweeds struck me as unusually fine. Thus
the dulse, (*Rhodymenia palmata*,) a species common
on our coasts, and eaten by the poor in Scotland and
Ireland, was fringing the sides of the pools, its broad,
deeply cleft, dark-red fronds, developed in great luxuriance. There were also large and dense tufts of
Chondrus crispus, the Irish or Carrigeen moss, as it
is called, when dried and sold in the shops, to make
jellies, for use in cookery, and for many other purposes. This, too, is a common species, and one that
varies much with the locality where it grows. When
found in shallow pools, considerably above low-water
mark, it degenerates in size, becomes of a pale olive
tint, and quite devoid of beauty. But see it at a
lower level, growing in some deep shadowy pool, as I
saw it here, and you would hardly believe it to be the
same. The fronds form large, bushy, and well-grown

tufts, with the leaves clean and glossy, and of a dark-purple hue; but what gives it its peculiar beauty is, that every segment of its many-cleft leaves reflects the most refulgent hues of azure and steel-blue. These tints, however, depend entirely on the submersion of the plant; remove it from the water, and every trace of them has vanished; replace it, and they as instantly reappear.

Another curious seaweed was *Codium tomentosum.* It forms thick cylindrical stems, much branched, and of a dark-green colour. Its appearance is downy, and, when touched, it has a soft spongy feel, and is enveloped in a slimy jelly. This curious plant was growing numerously here, imparting a somewhat singular aspect to the shallow pools, from the green velvet patches of its expanded bases, as well as from the stems.

The great tangles and oar-weeds were abundant, as were the sea-thongs already mentioned; and among them grew a much less common species, at least on the English shores, the henware, (*Alaria esculenta*), a large plant, much resembling the oar-weed, but of paler colour, and distinguished from it by having a stout midrib running through the whole length of the leaf. This midrib is eaten by the poor of our northern coasts, and of other parts of Europe.

Of marine animals I did not see many. The commonest species of sea-anemone (*Actinia mesembry-anthemum*) was speckling the rocks in its many varieties, for it is a very variable species, sometimes

chocolate-brown, or of all shades between that and a
glowing red. More rarely, it is dark olive, merging
into grass-green ; and not unfrequently specimens are
found, especially such as are of very large size, in which
both of these hues are combined, the ground colour
being dark-red, studded all over with small green
spots. This is the best known of all our native species ; indeed, it is the only one ever seen by thousands
who fancy themselves familiar with our sea-anemones.
The reason is not only the great abundance of this kind,
but its habit of living within tide-marks ; for such is its
patience of exposure to the air, that it may frequently
be seen sticking to rocks, particularly if shaded from
the sun, not far below high-water mark, where it must
be necessarily exposed to the air for many hours out of
every tide. Handsome as its appearance is, whether
displaying its smooth and glossy coat, or expanding
its crown of tentacles like a full-blown crimson flower,
it is the least beautiful, perhaps of all, and is not
worthy to be compared for beauty with some other
species which frequently dwell in its immediate neighbourhood, but in so retired a manner, that few, except
the professed naturalist, ever have the opportunity of
admiring their charms ; like modest worth, whose
excellence is often unknown or unappreciated, because
of that retiring humility which is its greatest grace,
while inferior pretensions are honoured, because they
are flaunted in the face of day.

In one of the crevices within the cavern I had
noticed a specimen of a far nobler species, certainly

the most imposing, if not the most beautiful, of all
the British sea-anemones, *Tealia crassicornis*. When
contracted, its body is usually of a rich crimson or
fine scarlet hue, often streaked irregularly with green,
like a ripe apple. Instead of being soft and glossy
like *A. mesembryanthemum*, it is hard and firm to the
feel, almost like leather, and its whole surface is rough
with numerous warts. It does not adhere to the exposed sides of rocks, but hides itself in dark holes and
narrow fissures. Nor is it satisfied with this protection, but for further concealment it covers its body
with a coating of gravel. This it does by means of
its warts, which are the terminations of so many
tubes, and which act as suckers, each one firmly
attaching to itself a small pebble or fragment of
gravel. When the animal is dislodged from its fortress, an operation by no means easy, and deposited
in a capacious vessel of sea-water, it presently throws
off the gravel, bit by bit, and stands revealed in all
its beauty, as if it were aware that its usual artifice
would avail for its concealment no longer. Soon,
however, it assumes a new form and greater magnificence. It expands a disk three inches in diameter,
fringed with many rows of thick conical tentacles.
These are of different colours in different individuals,
sometimes clear pellucid crimson, at others purple,
always surrounded with a broad ring of white.
Another variety of very charming appearance has the
tentacles entirely cream-white. The animal has the
habit of imbibing water, until all the tissues of the

E

body, as well as the tentacles, are filled with it, and swollen to a surprising extent. All the rich colours, especially those of the tentacles, are softened, diluted, and rendered translucent by this process; and the gorgeous array exhibited by a finely-coloured individual, when in this condition, can hardly be surpassed by anything of the kind.

With much fatigue and difficulty we made our way up the lofty slope, not altogether without danger, from the loose stones which the climbers were perpetually dislodging from the rubble, and rolling down upon the heads of those coming up below. Arrived at Benjamin's Chair, we sat a few moments to recruit ourselves, while our friend entertained us with anecdotes illustrative of the habits of the seal.

" I was one day standing," said he, " here at Benjamin's Chair, when I saw in the water below, which was clear and smooth, a large seal come up to the surface, carrying in his mouth a conger-eel, perhaps some eight or ten feet long, and as thick as my leg. The animal played with his prey, exactly as you have seen a cat play with a mouse; letting it go, then darting after it as it sought to escape, and catching it with perfect ease. All its motions were full of grace. At length the seal bit the fish in sunder with one snap, and, allowing one portion to sink, ne ate from the other till he reached the head. This he rejected, throwing it from him; then dived for the tail, which he brought up, and ate that in like manner.

" On another occasion, near the same spot, I observed a seal treating a salmon, which he had caught, after a similar fashion. It was astonishing to see how utterly powerless were all the attempts of the salmon to escape before the rushing pounce of the seal; it was overtaken and seized in an instant. When he was tired of his play, he suddenly tore off a large portion from the fish's side, and I assure you that the severing of the muscles was distinctly audible where I stood. In this instance the creature devoured the back part first, and, like an epicure as he was, reserved the belly for the *bonne bouche*.

"I believe our species is the common spotted seal *(Phoca vitulina)*; I do not think we have any other."

As we were returning, we made a slight deviation from our way, to see a hole which had just been discovered, and which was the present wonder of the little island's population. One of the men had noticed, in a particular part of the moor, that the earth returned a hollow sound. On digging, a block of granite was found a little below the surface. It was about eighteen inches thick, and was estimated to weigh five tons; its ends rested on two upright slabs, between which was a cavity, some six feet deep and as many wide. It was evident that the excavation had been made, and the stones placed, by human labour; and the latter operation must have been one of no small difficulty, from the great weight of the slabs; but for what purpose it could have been made, whether as a place for temporary retirement,

for some one who feared an enemy whom he dared
not resist, or for the secretion of valuable property in
some of the troublous times, of which the island has
seen many, there was no clue to inform us. No sub-
terranean passage was observed, though the earth at
one side was so loose as to suggest the notion that
such a communication might once have existed; a
fragment of pottery was the only object found. I was
myself struck with a rank odour in the cavity, very
different from that of newly-turned soil; the earth,
too, at one end, was black, and of an unctuous appearance, somewhat like that of a grave; but no trace of
bone or other organised matter could be found.

The appearance of this rude structure somewhat
resembled that of the monument known as Wayland
Smith's Cave, near Ashdown, in Berkshire. This
consists of a broad slab laid horizontally on several
upright ones. The earth in the lapse of centuries
had accumulated, until it was level with the flat slab;
but the lord of the manor, about thirty years ago,
cleared away the ground both within and without the
edifice. Local tradition assigns it to an invisible
blacksmith, who was said to shoe travellers' horses
there for a small fee. The money was to be laid on
a stone, and the steed tied; in the morning the
money was gone, and the horse was found shod. The
prescribed fee was sixpence, and neither more nor
less would do. Sir Walter Scott, in a note to " Kenil-
worth," suggests that this legend may have alluded to
" the northern Duergar, who resided in the rocks, and

were cunning workers in steel and iron ;" for there is little doubt that the monument is an accessory of the pile raised over the tomb of Baereg, the Danish chieftain, slain here in a great battle with our King Alfred. It is possible that the construction, the opening of which we saw at Lundy, may have an antiquity as great as its counterpart in Berkshire, or perhaps even greater, seeing that the huge upper slab was here quite covered with the common mould ; and, in default of any evidence to the contrary, we may conjecturally assign to it a similar commemorative purpose.

The next day was to find us upon the sea. Captain Tom Lee was going out to haul his pots, and we were to avail ourselves of the opportunity of becoming personally familiar with the vagaries of lobstercatching. A worthy fellow is Captain Tom ; kindhearted and obliging, one that has read a good deal, and has seen somewhat of the world, and free in communicating the knowledge he has acquired. We found him to be quite an agreeable companion, when he favoured us with his society. He unfortunately lost his ship on the African coast not long ago ; and since that time he has devoted himself to the fisheries of the island, which he prosecutes with energy and success. Captain Tom has been an attentive observer of the habits of animals. One anecdote of his was so good, that I think it worth preserving. But the captain shall be his own narrator :—

"A curious animal is a pig, gentlemen ! Very

cunning, too; a great deal more sensible than people give him credit for. I had a pig aboard my ship that was too knowing by half. All hands were fond of him, and there was not one on board that would have seen him injured. There was a dog on board, too, and the pig and he were capital friends; they ate out of the same plate, walked about the decks together, and would lie down side by side under the bulwarks in the sun. The only thing they ever quarrelled about was lodging.

" The dog, you see, sir, had got a kennel for himself, the pig had nothing of the sort: we did not think he needed one; but he had notions of his own upon that matter. Why should Toby be better housed of a wet night than he? Well, sir, he had somehow got into his head that possession was nine points of the law; and though Toby tried to shew him the rights of the question, he was so pig-headed that he either would not or could not understand. So every night it came to be . ' catch as catch can.' If the dog got in first, he shewed his teeth, and the other had to lie under the boat, or to find the softest plank where he could; if the pig was found in possession, the dog could not turn him out, but looked out for his revenge next time.

" One evening, gentlemen, it had been blowing hard all day, and I had just ordered close-reefed topsails, for the gale was increasing, and there was a good deal of sea running, and it was coming on to be wet; in short, I said to myself, as I called down the companion ladder for the boy to bring up my pea-jacket, 'We are going to have a dirty night.'

"The pig was slipping and tumbling about the decks, for the ship lay over so much with the breeze, being close-hauled, that he could not keep his hoofs. At last he thought he would go and secure his berth for the night, though it wanted a good bit to dusk. But lo, and behold! Toby had been of the same mind, and there he was snugly housed. 'Umph! umph!' says Piggy, as he turned and looked up at the black sky to windward; but Toby did not offer to move. At last the pig seemed to give it up, and took a turn or two, as if he was making up his mind which was the warmest corner. Presently he trudges over to the lee scuppers, where the tin plate was lying that they ate their cold 'tatoes off. He takes up the plate in his mouth, and carries it to a part of the deck where the dog could see it, but some way from the kennel. Then, turning his tail towards the dog, he begins to act as if he was eating out of the plate, making it rattle, and munching with his mouth pretty loud.

"'What!' thinks Toby, 'has Piggy got victuals there?' And he pricked up his ears, and looked out towards the place, making a little whining. 'Champ! champ!' goes the pig, taking not the least notice of the dog; and down goes his mouth again to the plate. Toby couldn't stand that any longer; victuals, and he not there! Out he runs, and comes up in front of the pig, with his mouth watering, and pushes

his cold nose into the empty plate. Like a shot,
gentlemen, the pig turned tail, and was snug in the
kennel before Toby well knew whether there was any
meat or not in the plate."

"Capital!" we all exclaimed; and so no doubt will
my readers exclaim, since the narrative may certainly
be relied on as authentic. I give it you as it was
told to us; and I am sure Captain Tom is too vera-
cious a man to invent or exaggerate the story.

The morning was foggy and unpromising, but the
prospect of lobster catching overcame the dishearten-
ing effect of the mist, and we were all upon the beach
in pretty good time and in pretty good spirits. When
we were at the water's edge the fog had lifted, not
resting upon the water, but with a thin stratum of
clear air between, through which we could discern
the surface of the sea to a considerable distance beneath the fog, which still filled all the higher air, and
enveloped all the land in a dense cloud. The massive
headlands, progressively receding into the distance,
loomed through the gray mist with fine effect, their
grandeur heightened by the indefiniteness which they
derived from their cloudy veil. We thought of some
of the effects in Turner's pictures.

The boat was moored some distance off-shore, and
we were indebted to the kindness of a brother fisherman, whom our worthy skipper hailed, for putting us
on board in his punt. Here, then, we were embarked,
—Captain Tom and his man Dick, and we three
idlers. Scarcely a breath of wind was stirring, and

the misty air fell heavy and cold, but we pulled along inshore with hearty good will. The cormorants and gulls swept by us, wondering at the intrusion; the former, with outstretched neck and flapping wings, flying in straight lines, as if with some definite point in view, just as men of business press along Cheapside or Mincing Lane; the latter on easy graceful wing, sweeping round in circles, as if intent only on amusement, as ladies stroll in the parks. Presently came flying by two oyster-catchers, or, as the men call them, sea-pies, conspicuous in black and white plumage, and with beaks and feet as brilliant as red sealing-wax.

We passed some fine caverns in the cliffs, and on the points of rock far above were seen two or three of the wild goats, of which there is a flock on the island. It was amusing to observe with what fearless ease and precision of footstep they jumped and scampered about the peaks, delighting to come to the very verge of the precipice, and to run along the ledges not more than a few inches wide, or to stand upon the tottering masses, and gaze down upon the sea.

When we came opposite the half-way wall, where the granite takes the form of ancient masonry,—so that one can scarcely help imagining that the cliffs are crowned with the remains of walls and towers, built by fabled giants of the olden time,—we began to find ourselves once more in the midst of a dense population of birds. There were plenty of guillemots, speckling the gray rock with their dusky forms in rows of black dots. Their numbers appeared to render sitting-space an object worthy of contention; for whenever any of the flying squadron attempted to land, and to intrude himself among his resting fellows, he was invariably met with opened wings and beaks, and the most threatening demonstrations of resistance, like Cæsar when he landed on our shores from Gaul. But the characteristic bird here was the kittywake, or hacklet, a very small species of gull, with the upper plumage of a delicate French-gray hue, and the lower parts white. They also sat in rows on the narrow shelves, each one with a nest of dried grass beneath it, like so many Turks in a mosque, squatting each upon his own bit of carpet. Their size, form, and colour gave them the closest resemblance to doves,— a resemblance which was not a little increased by some traits of their manners. Two sitting next each other would occasionally bring their beaks together in that playful toying manner which every one must have seen our common pigeons practise, and which is so much like kissing, that it is hard to imagine it any other than an expression of affection. It was suggested that one was feeding the other, but I am rather disposed to put the former interpretation on the action. The common name of this little bird is derived from its cry, " Kittywake, kitty-kittywake;" but the sounds as correctly express the words, " Get away, get away," which we took as a polite intimation on the part of the birds that our morning call was an unseasonable intrusion. We clapped our hands smartly, and the air was instantly filled with birds, though many of the sitters held fast to their nests. The guillemots flew out to sea, but the kittywakes, after a turn or two, in which their little black feet contrasted curiously with their snowy plumage, returned to pursue their domestic occupation.

We had lain upon our oars for a few minutes to gaze upon the birds, but time was going, and we had other fish to fry. The men accordingly gave way, and as the boat shot off, the little gulls, as if in joy, could not refrain from hastening our departure with renewed vociferations, which rose at the same moment from every ledge, as if by common consent, of " Get away! get away!"

Near this part the cliffs become much lower than usual. Here, in the time of Charles II., a fort was erected, which was furnished with brass cannon. Local tradition commemorates this circumstance in the title of "The Brazen Ward," still applied to this point; and the old brass guns themselves are said to be visible in calm weather and clear water, far down in the depths, whither they were thrown overboard by the French when the fort was dismantled. This event took place in the reign of William III. The stratagem by which the unscrupulous Frenchmen got footing on our island, which might well have been deemed impregnable, is curious as illustrating the usages of war.

The island at that time was more extensively cultivated than at present, and supported a population more than twice as numerous. Barley, potatoes, and

all kinds of culinary vegetables were raised in great
abundance; the fields were well stocked with cattle,
sheep, and goats; a brisk trade was carried on in the
skins of rabbits, which then, as now, perforated the
barren slopes by myriads; and the resources of the
inhabitants were increased by the sale of feathers and
eggs, the produce of the sea-fowl which every summer
tenanted their cliffs.

Confiding in the natural strength of their insular
rock, the inhabitants dwelt in unsuspecting security,
notwithstanding the war that raged abroad. One day
an armed ship was seen to anchor in the roads. She
hoisted the national flag of Holland, with which country England was at that time in amity, and presently
a boat was seen to leave her side and pull for the
landing. The crew, in imperfect English, contrived
to make themselves understood. They stated that
they had mistaken the proper channel, and had taken
shelter in the road; that their captain was lying grie-
vously ill, and that supplies of milk, and other little
luxuries of that kind, would be a desirable addition to
his comfort, and would be gratefully received. The
simple people believed the story, and readily granted
such supplies as were desired, which were regularly
fetched for several days in succession. At length the
crew reported that their captain was dead, and they
requested, as the last favour, that if there were any
church or consecrated ground on the island, they
might be permitted to deposit the corpse in it; and
they intimated also, that it would be an additional

favour if the principal persons of the island would be present at the burial. Everything was promised without suspicion; and the greater part of the inhabitants, arrayed in their best garments, assembled to render the last honours to the foreigner by following his body to the grave. They even volunteered their assistance to carry the corpse, as the chapel was more than a mile distant on the other side of the island, and the access to it was not, as now, by a good broad road, but by steep and difficult paths. The coffin, indeed, seemed more than usually heavy; but they supposed that the deceased captain might have been a very corpulent man, especially as Dutchmen are reputed to manifest a tendency to a somewhat bulky build, and, therefore, this circumstance passed without exciting any particular notice.

The little chapel is at length reached, the corpse deposited on a bier, and the burial service commenced. A little hesitation occurs; one or two of the foreigners whisper among themselves; and then one of them steps up to the islanders, respectfully intimating that the customs of their religion forbid those of a different persuasion to be present at that part of the ceremony which is now about to be performed. It will, however, he assures them, occupy but a few moments, after which they shall be readmitted to see the interment. The inhabitants comply with prompt courtesy, leaving the strangers in undisturbed possession of the chapel.

In a few minutes the door was thrown open, and a

band of armed men rushed out, who took their aston-
ished and unresisting hosts prisoners. The whole
had been a *ruse de guerre*, a vile and complicated
falsehood, with which the inhabitants, by their very
kindness and courtesy, had been beguiled to their
ruin. Instead of Dutchmen, they found that they
had to do with their wily and bitter enemies the
French; and learned, with unavailing regret, that
they had helped to carry upon their own shoulders, in
the coffin, those arms which were destined to make
them captives.

The whole of the island was now ravaged without
mercy; and, not content with robbing the poor people
of such portions of their property as could be carried
away, the invaders wantonly and wickedly destroyed
the remainder. The historians of the time state that
the island contained at this period fifty horses, nearly
the same number of neat cattle, three hundred goats,
and five hundred sheep. The greater part of the
horses and cattle they hamstrung, so as to disable
them for use, and the goats and sheep they threw over
the cliffs. They took away even the clothes of the
wretched inhabitants; and so bent were they on destruction, that a large quantity of meal happening to
be in certain lofts, under which was salt for curing
fish, they scuttled the floor, and so, by mixing the
meal and salt together, spoiled both. They then
went over to the fort on the eastern side, dismantled
it, threw the brass guns into the sea, as I have already

mentioned, and left the scene of their villanous exploit destitute and disconsolate.

A little way beyond the Brazen Ward, there is, on a projecting headland, a large square block of granite, with one end resting on a smaller piece, exactly in the same manner as a brick is tilted upon a bit of stick, to form a rude but effective trap for imprudent mice. The block rests on a smooth platform, and stands in dark relief against the sky, while just behind it there is a natural perforation in the rock, through which the light streams brightly. The Mouse-trap and Mouse-hole are the designations applied to these curious objects; and I thought them so interesting, that I begged to be put on shore for a few minutes to sketch them. The swell made landing and re-embarking rather a ticklish business; but I managed to effect both the one and the other without a wetting, and found myself on one of the narrow ledges, just above the water-line, where I made the accompanying drawing of the scene.*

We now approach Gannet Rock, that church-like mass of granite which I have before mentioned. It stands just in front of a projection of the coast, forming, with one of the points which we have just passed, a little bay, somewhat deeper than a semicircle. We understood that the Admiralty had contemplated to select this as the site of the Harbour of Refuge, which has so long been thought desirable on the internal

* See page 45.

side of the island. It is supposed that a comparatively little outlay would effect the purpose here, as all that is necessary is to fill the interval between Gannet Rock and the Point, and to form a pier or breakwater from the outside of the former, so as to narrow the entrance to the Cove.

But by this time we had commenced the business which had been prescribed as the chief object of the excursion. All along this end of the island is excellent ground for lobsters, and here Captain Tom had sunk some thirty or more of his pots. These were in succession hauled up and examined. They are set at considerable distances apart, and the place of each is indicated by buoys of cork, affixed at certain intervals to the rope. But it was now spring-tide, and the time of high-water was scarcely passed ; hence some of the buoys were submerged, their length of rope being insufficient for the depth of water. The position of these, therefore, could not be determined ; and though the captain and his man knew by the bearings of the land whereabout to look out for each, they had to wait for the successive " watching" of each buoy, as its first appearance on the surface is technically termed, before they could haul.

The form of a lobster-pot is generally known, as there are few of our rocky shores where the simple but effective contrivance may not be often seen lying on the beach. Their principle is that of a wire mouse-trap ; they are made of strong osiers, with a rounded top, the points bent inwards at the centre, so as to allow of the entrance, but not of the escape, of the lobster.

Each pot, on being hauled to the surface, was pulled on board;· the next thing was to take out the prey, if any were there. These were of four different kinds— the lobster, the most valuable of them all; the sea cray-fish, or thorny lobster, larger, but in less estimation, the flesh being dry and somewhat hard; the common crab, the value of which is generally appreciated; and the spider-crab, or maia, of little value as food, though occasionally eaten.

It was interesting to notice the different habits of these species. The lobster was agile, but cool, and thoroughly prepared for war, holding up its large, formidable claws, widely gaping, in a reverted position over the back, so that it was rather a dangerous affair to get hold of one. The expertness acquired by practice, however, enabled the fisherman to dash his hand through the entrance of the pot upon the animal's back at the fitting moment, and suddenly to drag him up stern-foremost.

The cray-fish, active, but large and unwieldy, seemed conscious that he had no powers of defence to be compared with those of his cousin. The claws in this species are small and feeble; but, equally unwilling to be made a prisoner, he endeavoured by agility to supply the lack of weapons; flapping round and round the circle of the pot, by means of rapid and forcible blows with his expanded tail. We noticed the singular sound produced by this animal when

F

excited; the bases of the antennæ are studded, as is indeed the whole surface of the animal, with prickles; and these it rubs with force against the sides of the shelly horn that projects from the forehead, by which a singular grating noise is made, accompanied with a very perceptible vibration. Our friend the captain, who has the misfortune to be deaf, protested that he could hear the sound distinctly whenever he touched the animal with his hand; but I am not sure whether this was not a confusion of senses; a mistaking of the vibration of which his nerves of touch were cognizant, for such as would have been appreciable by those of hearing.

The crabs, on the other hand, both the common kind and the spider, were sluggish, inert, and helpless, yet somewhat awkward to take hold of, and to pull out of the entrance, on account of their breadth. The spiders, too, like the cray-fish, are bristled over with stout, sharp-pointed spines. The contrast between the agile power of the lobster and the torpidity of the crab, when taken from their proper element, is very striking. The former, as I have said, presents his threatening claws to his adversary, like a warrior skilled in the use of his weapons and prepared to use them; leaping and springing about, at the same time, with a sort of dashing recklessness, as hoping to find some possibility of escape, even from the worst circum-stances. The crab seems paralysed as soon as he is taken out of the water. Though furnished with claws of a stony hardness, apparently superior in the power of grasping and pinching to those of his nimble cousin, he rarely attempts to use them; but folding them together, and crumpling up his legs stiffly across his breast, he is content to lie passive, and abide his fate. You may take him up in your hand, turn him over, and examine him; not a limb will he move; nay, you may even put him in your coat-pocket, and carry him for a mile, and, on taking him out, find him as patiently resigned as when you put him in.

As soon as the captives were secured, any pieces of old bait that remained were shaken out into the boat, to the no great delectation of our olfactories. This was destined to be thrown overboard, but not *here* upon the lobster-ground, lest it should interfere with the temptation of the traps. Fresh bait was now introduced: the fisherman, taking a piece of skate about as big as his hand, pierced a hole through it with a marling-spike, to receive a wooden skewer, pointed at one end and cut in a peculiar manner, with a sort of shoulder in the middle. The skewer, thus baited, was put through the side of the pot, and the point being inserted between the close-set osiers of the mouth, it was then tightly driven in with a stone. By this contrivance the bait is fixed within the trap at such a height as prevents the captives from getting at it readily, while it cannot be reached from without.

The peculiarly rough surface of the spider-crab renders its shell a suitable *nidus* for the growth of parasitic plants and animals; and I think we did not take an individual that was not studded more or less densely with zoophytes of the genera *Sertularia*, *Plumularia*, &c., sponges, and sea-weeds. Some curious forms inhabiting the deep sea are occasionally in this manner presented to the observant inquirer, which he would otherwise obtain only by means of the dredge.

The course of our examination of the successive lobster-pots had by this time brought us to the north-east point of the island. All the buoys had not yet "watched;" but there was here a tremendous sea running, and the swell kept setting us on the rocks so fast, that not only we landsmen, but even the fisher-men, began to doubt the prudence of remaining in a situation so exposed any longer. Add to this, that heavy thunder-showers had already drenched us to the skin; we were thoroughly cold, and our limbs were cramped from sitting for hours in the stern-sheets of the narrow boat. It was, therefore, not without inward satisfaction that we heard our friend Tom decide to give up the remaining pots, and make the best of our way into smoother water.

LUNDY ISLAND—(*concluded.*)

ENTRANCE TO THE CAVERN.

We had contemplated another pleasure to be included in this little trip, which we felt reluctant to relinquish, although the rain by this time had begun to come down in that settled, steady manner, which makes you feel that it intends to do business for many hours to come. The men had put an oyster-dredge into the boat; and I for one looked forward with interest to this essay in rifling the treasures of the deep sea.

On the eastern side of the island the proprietor, some years since, had endeavoured to form an oyster-bed: the ground was suitable, and he had stocked it with living oysters. The result of the experiment had not as yet been tested, and it was proposed that we should make the first examination.

The dredge, as most of my readers are probably aware, is a bag attached to an iron frame, one side of which is bent outward, so as to form a sort of lip or edge, for the purpose of scraping the ground. The lower side of the bag, or that which drags over the bottom, is formed, not of any textile materials, but of large iron rings, interlocked so as to make a loose chain-work. To a bridle across the mouth a rope is attached, of sufficient length to allow the dredge to lie on the bottom at a considerable distance astern. In action, the dredge is dropped overboard carefully, so that it shall fall lip downward; the rope is allowed to run out to a sufficient length, and is then passed over the stern, and belayed. The boat is now rowed, or sailed, if the wind be fair, over the ground; and its motion being communicated to the dredge, the iron lip scrapes up and lodges in the bag whatever lies loose upon the bottom. The mud, sand, and shingle, which are scraped up also, and everything, in short, that is much below the size of an oyster, passes through the iron meshes or links of the chain, while everything above their size is retained. After

a while, according to the judgment of the operator,
the dredge is hauled up, and the proceeds examined.
For this purpose the rope is shifted to the middle of
the boat, and the contents of the dredge are emptied
out.

Our success was not very encouraging. We made
three hauls, and brought up a few oysters, which
were tolerably good. Some of them were evidently
old fellows, so old that we conjectured that they might
possibly have been among the original fathers of the
colony. The rough and laminated shells of these
were studded with small seaweeds and zoophytes, and
several of those agile creatures, the brittlestars, were
sprawling their long flexible limbs, like so many snake-
tails, over their surfaces. Some of the zoophytes I
preserved for microscopical examination when I should
arrive at home; and their elegant forms and curious
structure well repaid the observation.

Among them was the beautiful *Plumularia Catha-
rina*. This zoophyte, which may be taken as the
representative of an extensive family, grows up like a
tiny plant, having a single stem, with many branches,
like a miniature tree, or many stems, springing up
in a tuft or cluster, like a shrub. Both stems and
branches are composed of transparent horny tubes,
forming false joints at frequent intervals, and developing at various points little shallow cups. This is
the skeleton. Every part of the tubular stem and
branches is permeated by a fleshy core or pith, which
in every one of the little cups develops itself into a

polype, having many highly-sensitive tentacles, which
expand like the rays of a star around the mouth.
When in health, and undisturbed, these exquisite
organs are stretched in all directions, resembling so
many threads of spun glass ; but on the slightest
touch, or even on a shock being given to the vessel
in which the animal is kept, the tentacles contract
into shrivelled and shapeless lumps, and the whole
animal shrinks down to the bottom of its cup-like cell.

Another of the plant-like forms of compound life,
but belonging to a class of higher organic rank, was
Crisia eburnea, called by Ellis the Tufted Ivory Coralline, an appellation which well indicates three of
its prominent qualities; its stony coralline texture,
its delicate whiteness, and its habit of growth in little
bushy tufts, about an inch in height. The cells here
are short tubes, and the polypes, which project from
them, have a much higher organisation, a more complex form, and more precise and energetic motions,
than those of the *Plumularia*. The tentacles in this
species are not contractile in their own substance, but
are capable of being closed together in a parallel
bundle, and of being withdrawn into the body, as
into a sheath. They are again expanded by the
turning inside-out of the integuments which sheathe
them, just as a stocking or a glove is reversed.

The substance of the skeleton in the class of animals to which the *Crisia* belongs, is composed of
lime ; hence it is brittle, and of a stony hardness. If
a small portion be held to the flame of a candle, there

will appear, at the very edge of the flame, a light of
most intense brilliancy, which is but another exhibi-
tion of the principle on which is produced the celebrated *lime light*, recently brought into notice for its
superior power of public illumination. The whole of
the substance of the cells, when viewed through a
microscope, is seen to contain a number of clear oval
grains, very much like the bubbles which we occasionally see in bad glass; they are, however, regular
in size and in arrangement. Their nature and use
are, I believe, entirely unknown.

Through one of these oysters I made acquaintance
with another form of the same class, which has more
the appearance of a membranous seaweed than an
animal, the Bugle coralline (*Salicornaria farcimi-
noides*). It forms many slender flattened branches,
swelling regularly between the joints, and covered all
over their surface with ridges or raised lines, set
diamond-wise, and enclosing depressed cells of the
same form. The polypes which inhabit these cells are
probably similar in form to those of the *Crisia;* but
I could not detect a single individual on the specimen
that I examined, and I know nothing of them.

Upon the whole the excursion of this day, though
accompanied with some unpleasant circumstances,
from the state of the weather and the sea, was one of
much gratification. The disagreeables were nothing,
or at least they lost their disagreeable character, as
soon as they had actually ceased; while the pleasure-
able emotions produced upon the mind were repeated

as often and as long as memory dwelt upon them. For the memory of pain is not painful, while the memory of pleasure is often little less pleasant than the first enjoyment of it.

Dining with the hospitable proprietor, we gleaned some fragments of information on the natural history of this little isle, that we should have had no opportunity of learning by actual observation. The boggy moors in the elevated centre of the island afford a suitable rendezvous to the woodcock and the snipe; and sporting gentlemen occasionally come over, expressly to take the former on their first arrival, which usually precedes their appearance on the mainland by several days. Swallows and swifts we should expect to find here; but I was somewhat surprised to learn that the goatsucker is a regular summer visitor, as we commonly associate this bird with groves and woods, of which the isle is absolutely deprived. Among the occasional visitants were mentioned the rose pastor and the hoopoe, both birds of considerable size and of great beauty. The wild duck, the widgeon, and the teal, are sufficiently numerous to afford first-rate sport. The peregrine falcon breeds in the lofty cliffs, especially in those of the exterior side. One of the farm labourers shewed me a pair of well-grown birds which he had reared from the nest; they were in excellent health and condition, and in full plumage. The nest had been rifled by a boy let down from above for the purpose, at that part of the perpendicular cliffs which is immediately over the Seal

Cavern. The fellow was in the habit of feeding his pets with the flesh of the puffins and guillemots, which his dog would catch for him in any desired quantity. The osprey, though less common than the peregrine, is not unfrequently seen fishing around the rocks.

Of small birds, the chaffinch and the linnet are common; but, what is strange, the sparrow is not found. The song-thrush is a constant resident, which finds its favourite food in the pretty banded hedge snail (*Helix nemoralis*) that is also common. The song of the skylark we had heard saluting the sun on each of the brilliant mornings that we had spent on the island; and the pipit was hopping and flitting about the rocks all round the coast.

We had already noticed many *insects*, but were hardly prepared to hear that an entomological gentleman, well known to us by reputation, had recently obtained, during a visit of only a few days to the island, more than three hundred species, the great majority of which were beetles.

The scenery of the western side is more magnificent than that of the eastern. The precipices generally attain a more stupendous height, and the prospect seaward is an entire semicircle of unbounded water, expanding to an immense width. Alternate indentations and projections in the line of coast, shallow coves and lofty promontories, occur all along; and as the visitor wanders by the margin of the cliff, he is continually charmed by newly-opening and ever-changing views of the shore, ever-fresh combinations of the

massive granite rock, and resemblances the most close
to vast works of human art.

One of these promontories appeared to me peculiarly grand, and tempted me to spend an hour in
endeavouring to convey with the pencil somewhat of
its character, though with only partial success.* There
was a cavern cut, as it were, in the nearly perpendicular stone, of great height, but comparatively narrow,
and with the sides so nearly parallel and straight, that
it looked like a gallery or passage built with cyclopean masonry; while the massy abutments on each
side were so symmetrical, sloping upward from broad
pedestals, that I could almost have fancied them the
enormous *propylea* of some old Egyptian temple, the
stones of which were partially disjointed and disintegrated by the wear of four thousand years. The
surf was boiling and beating without, rearing itself in
futile rage against the foot of the promontory, only to
be ever driven back upon itself, like brave warriors
vainly assaulting the impregnable walls of some
mighty fortress; or, as the poet has expressed it,—

"Wrestling with rocky giants o'er the main,
Which spurn'd in columns back the baffled spray."

Within the cavernous gallery the water was smooth
and glassy, rising and sinking indeed with ceaseless
undulation as the wave rose and fell, but reflecting as
from a surface of polished steel the blackness of the
obscure interior. The utter solitude of the scene increased its grandeur; no trace of man or his works,

* See engraving on page 85.

no hut, no fisherman's net, no boat, not even a distant ship, broke in upon the majesty of nature; and though thousands of sea-fowl were playing about the point, or sitting in crowded rows upon the steps and pedestals, their distance reduced them to mere specks so minute as scarcely to be obvious to sense, and did not affect the general impression of loneliness.

Oh! it was beautiful to sit in the bright morning in the deep quietude of these heath-covered heights, and gaze down upon the glorious sea! To get under the shadow of one of the mighty blocks, squared almost as with the stone-hewer's chisel, that crown, as if with ancient ruined fanes, every projecting headland, and there enjoy the beauty and the exhilaration of the sunlight, without feeling its oppression! And how rich and glorious is the flood of light that bathes every object in the unclouded sun of summer! How full and deep the shadows, how broad the lights, on such a broken coast as this! How rich and lovely the colouring of blossom-sheeted heath, expanded sea, and vaulted sky! "Truly the light is sweet, and a pleasant thing it is to behold the sun." What heart cannot respond to the exquisite stanzas of one who drew her inspiration from the grandest and most majestic scenes in nature? Who cannot sing her passionate lay "To the Sunbeam?"

> "Thou art no lingerer in monarch's hall;
> A joy thou art, and a wealth to all;
> A bearer of hope unto land and sea.
> Sunbeam! what gift hath the world like thee?

" Thou art walking the billows, and Ocean smiles;
  Thou hast touch'd with glory his thousand isles;
  Thou hast lit up the ships, and the feathery foam,
  And gladden'd the sailor, like words from home.

" From the inmost depths of the forest shades,
  Thou art walking on through their green arcades;
  And the quivering leaves that have caught thy glow
  Like fire-flies glance to the pools below.

" I look'd on the mountains,—a vapour lay
  Folding their heights in its dark array;
  Thou brakest forth, and the mist became
  A crown, and a mantle, of living flame.

" I look'd on the peasant's lowly cot,
  And a something of gloom enveloped the spot;
  But a gleam of thee on its lattice fell,
  And it laugh'd into beauty at that bright spell.

" To the earth's wild places a guest thou art,
  Flushing its waste like the rose's heart;
  And thou scornest not from thy pomp to shed
  A tender smile on the ruin's head.

" Thou tak'st through the dim church-aisle thy way,
  And its pillars from darkness flash forth to day;
  And its high pale tombs, and its trophies old,
  Seem bathed in a flood as of molten gold.

" And thou turnest not from the humblest grave,
  Where a flower to the sighing winds may wave;
  Thou scatterest its gloom, like the dreams of rest;
  Thou sleepest in love on its grassy breast.

" Sunbeam of summer, O what is like thee?
  Hope of the wilderness, joy of the sea!
  One thing is like thee, to mortals given,
  The faith touching all things with hues of heaven!"

                             HEMANS.

The chief curiosity of this side of the island is what is familiarly known to the inhabitants as the " Earthquake." It is a chasm, evidently the result of a great convulsion of nature; and local tradition confidently assigns it to that tremendous shock, in 1755, in which Lisbon was overwhelmed, and which was felt over nearly the whole of Europe. The ascription to it of such an origin has been ridiculed, but on very insufficient grounds. No one, I think, can look upon it without feeling the conviction that it has been produced by an earthquake; and the one to which it is currently assigned is to the full as likely to be the true one as any other.

We were directed to pursue the coast-line, along the edge of the cliffs, until we should reach the middle of the island, nor was there any difficulty in finding it, or in recognising it when found. It is a yawning chasm, or cleft, in the granite, running along in a line irregularly parallel to that of the precipice, for about five hundred feet. The width varies in different parts, but may be taken at fifteen feet upon the average. The sides of the cleft are quite perpendicular, to a depth of fifty feet. They are fringed with luxuriant ferns, and the common flowering plants that grow upon the sea-cliffs. The whole ground and rock round about, for some distance, is much shaken, and broken into chasms and fissures.

There is a second smaller cleft, which I had well-nigh overlooked, though it is, in fact, the more interesting of the two. It is situated much nearer to the edge of the cliff, and goes down to a depth nearly double that of the former. The rocky sides, which are from three to six feet apart, are very plane and parallel, yet slightly approaching as they descend. We were able to scramble down to some depth in the narrow fissure, and to obtain a glimpse, through slender cracks and crevices, into cavities apparently large, but unconnected with the air, and utterly dark. They gave forcible intimations, however, that the tearing of the solid granite rock had been much more extensive than one would suppose from merely viewing the superficial chasms.

A short time ago a large and beautiful amethyst was discovered imbedded in the rock some distance down, partially exposed by the cleft, in the line of which it happened to lie. The proprietor, who had himself made the discovery, and who thus possessed a double claim to it, wished to obtain the aid of a professed lapidary in extracting it; but, meanwhile, some greedy and dishonest person, who had got wind of the discovery, endeavoured to secure possession of the prize. The unskilful hands and clumsy tools employed managed, indeed, to deprive the rightful owner of the gem, but with no advantage to the covetous plunderer. In the rude efforts to extract it, the beautiful crystal became split and crushed to worthless fragments. We saw the hole which the rough chisel had produced, and the remains of the lovely gem still partly embedded in the stone, but beaten and pounded to a purple dust. A much smaller specimen was subsequently discovered near the former, and this was extracted without injury. Its value, however, was far inferior to that which the former would have possessed.

In the angles and crevices that occurred in the obscure walls of the chasm, I found several colonies of that curious insect the seaside bristletail (*Machilis maritima*). It is interesting to observe the brilliant refulgence of metallic colour bestowed on a creature nocturnal in its season of general activity, and haunting obscure recesses during the day. The insects of this genus are clothed with minute scales, whose edges lap over each other. In full-grown specimens of this species the scales reflect prismatic colours, undistinguishable, indeed, into individual rays, yet producing a combined effect of varied hues, very rich and lustrous. In many specimens, especially those of younger age, this colouring is much less conspicuous, or altogether lacking, being replaced by a dark iron-grey tint. The scales, taken singly, form beautiful microscopic objects; they bear the closest resemblance in form, structure, and markings, to those which cover the wings of butterflies, and to which all the varied hues and patterns of those lovely insects are owing.

We returned from the Earthquake through the Valley of the Punchbowl, the course of a little brook, which originates near the middle of the island, and forms there a pond of considerable expanse, and then winds, half-concealed, through a spongy bog to the edge of the cliffs. The smaller duckweed (*Lemna*

G

minor) was found partially covering the surface of the pool with a mantle of deceitful verdure; and one of the numerous kinds of pond-weed (*Potamogeton*) was floating in the brook, together with the water-crow-foot (*Ranunculus aquatilis*), a plant remarkable for the very diverse appearance assumed by its leaves under difference circumstances. It commonly grows in the midst of water; such of its leaves as reach to the surface, and are exposed to the air, are three-lobed and very slightly notched; while such as grow immersed in the water are cut into narrow threads, almost as fine as hair.

The mossy bog, which felt to the foot as if we were treading on a saturated sponge, yielded us two interesting plants. The one was the asphodel, (*Narthecium ossifragum*), a spike of small lily-like yellow flowers springing from a creeping root. The other was the sundew (*Drosera rotundifolia*), one of the few plants that form natural insect-traps. It was the first time that I had ever seen it in a living state, and I looked with much interest on its radiating crown of rounded leaves, each set at the end of a flattened foot-stalk, and covered with red hairs or bristles. On plucking a leaf to examine it more closely, we perceive that every one of these minute hairs is tipped with a globule of fluid, as clear as a dew-drop, but as clammy and adhesive as glue, capable of retaining small flies and other insects which incautiously alight upon the leaves. This viscous fluid is exhaled by glands at the extremities of the bristles, under the influence of the sun's rays, whence the common English appellation, as well as the scientific one, the word *Drosera* being derived from the Greek δρόσος, dew. That the object of the secretion is the capture of insects is highly probable, from what we learn by comparison of these with other plants, where a similar end is obtained by different means ; but of what benefit to the plant the prey can be, when captured and detained by so ingenious a device, botanists have not as yet been able to decide. It is conjectured that some element may be given out during the decomposition of the animal substance, which may be requisite for the sustenance, or at least the health, of these strange plants. Both the asphodel and the sundew were growing in considerable abundance in this particular locality.

We came now to the curious object which gives name to the little valley, the Punchbowl. It is a basin of the common granite, four feet in diameter, and one in depth, with a uniform thickness of six inches. Both the concave and the convex surfaces are segments of very perfect spheres ; and the whole conformation is so regular as scarcely to permit a doubt that it is the work of art. And yet, when we inquire what could be the purpose of such a piece of sculpture, and how it could have got to a situation so wild, so remote from any trace of man, and altogether so unlikely as the side of this boggy valley,—especially considering that its weight must have presented no small obstacle to its removal from any other

locality,—we know not what answer can be returned.
The only suggestion that appears at all probable to
my own mind, is, that it may have been the baptismal
font of some very ancient chapel, of which no other
vestige now remains. Even its hard and solid substance has begun to yield to the gnawing tooth of
time,—" tempus edax rerum ;" for the vicissitudes
of the seasons are already dissolving the bond which
united the heterogeneous materials of feldspar, mica,
and quartz, in one mass; and disintegrated nodules
of the last-named substance are lying loosely in the
concavity, as if a smart hail-storm had just expended
itself.

We could not leave the island without paying a
visit to the lighthouse. We had watched, evening
after evening, from the thronged promenade of Capstone Hill, its brilliant torch-like flame, as it appeared,
first a tiny spark, gradually increasing to a ruddy
glare, then waning to a spark again, and, after a few
seconds of total darkness, reappearing, to go through
a similar evolution.   Night after night, on those warm
dewy summer evenings, had we lingered on the rocks,
with scores of other idlers as interested as ourselves,
to mark the first appearance of the light on distant
Lundy, and, watch in hand, to count the moments
which, with unvarying regularity, elapsed between the
successive revolutions.

The lighthouse, which has been built rather more
than thirty years, is placed on the highest summit of
the island, a point not quite five hundred feet above

the level of the sea, but its own height elevates the lantern eighty feet above this. The white pillar-like structure is conspicuously visible from almost all parts of the island, and it often seems nearer than it really is. It looked but a very little way behind the Farm, but we found it the walk of a mile. Lapwings were wheeling round us with their well-known rapid circling flight, as we walked across the moor, uttering, sometimes close to our heads, and the next moment at a distance, their plaintive cries of " Peewit! peewit!"

The lighthouse appears a structure of great strength, built of massive hewn stones of granite, as well as the accessory buildings appropriated to the use of the lightkeepers. From the purity of the atmosphere on this lone rock, the whiteness of the stone is still unsullied by speck or stain; and the period of its duration is as yet too brief for the action of the weather to have had any perceptible influence in wearing down the angles of the stone, or even in defacing the lines of the quarryman's chisel.

A staircase of stone steps leads up to the lantern, which is a room fifteen feet in diameter, surrounded by panes of thick plate-glass about two and a half feet square. The light is placed in the centre, within a cage, having an octagonal revolving frame. Each of the eight squares of which it is composed, consists of many large lenses of varying powers, so arranged that the light shall be in the focus of all. In order to accomplish this, the central part of every lens, except the middle one, is cut away, and thus we behold a perfect lens in the centre, surrounded by successively diminishing segments of larger lenses. Square mirrors are placed both above and below, in many rows, at such angles as shall reflect the light upon the surface of the sea.

The whole combination of refraction and reflection has the effect of producing a most intense glare, when the eye of the beholder is immediately opposite the centre of any one of the lenses. The power of the light, indeed, may be imagined, from the fact that it shines with a strong and vivid glare at Ilfracombe, which is twenty-two miles distant. But this intensity of light is only momentary : by means of wheel-work, the motive power of which is a weight-and-chain pulley, like that of a clock, the eight-sided frame revolves around the light, with a uniform motion, performing the complete circle in sixteen minutes. Thus a period of two minutes elapses from one moment of greatest intensity to the next; the interval being occupied by a gradual diminution of the apparent light, till the dimmest point is attained ; and then a gradual increase to the brightest. At a great distance there occurs an interval of total obscurity ; but this is only because the rays are too feeble to be appreciable so far. Within a circle of a few miles the light never quite disappears.

The fatality which the lanterns of lighthouses occasion to birds has been often mentioned; it is, however, a curious circumstance. Lundy Light, it appears, is responsible for its full share of these casualties.

The keepers informed us that sometimes four dozen birds are found in a single morning, either killed or helpless, outside the lantern. They mentioned blackbirds as habitually flying against the panes, and fluttering down until they are caught in the gallery. Snipes dash against the glass with such force as to cut open their breasts ; a result, no doubt, promoted by the sharp and knife-like ridge of the breast-bone. Probably many of these accidents are attributable to the early habits of birds, wakeful and active before the glare of the artificial light has been dimmed by the advancing day; but, doubtless, many occur to migratory birds, performing their long aerial voyage ; as birds of passage are generally believed to perform their journey under cover of night.

I did not hear that these involuntary attacks had ever the effect of injuring the plate-glass against which they are directed ; but at the Eddystone Lighthouse it is recorded, that one of the panes was shivered to pieces by the forcible flight of a gull, to which it was no less fatal. The bird was found dead in the gallery, a pointed fragment of the glass, two inches in length, having penetrated its throat. The force of the shock was less a matter of surprise, when it was discovered to be that large and powerful species, the herring-gull.

So great is the power of the lenses, that, when the sun is shining, the keepers are compelled to exercise caution in entering the lantern for the purpose of cleaning the lamps. The concentrated rays would quickly set their clothes on fire, if brought into the focus ; blinds are therefore necessary, which are always kept down during sunshine.

The lamp is a large Argand burner, of four circular wicks, placed concentrically, or surrounding each other, with intervals between. In descending, we were shewn into a chamber filled with the large cylindrical glass chimneys to be used for the lamp : here they are kept in store, arranged on shelves round the room. Eighteen dozen, as we were told, was the number that we saw. The stores are replenished at certain intervals from a vessel loaded and sent round by the Trinity House, to visit in succession all the lighthouses on the coast.

At the bottom of the edifice there is a second light-chamber facing the sea. Here are placed nine hemispherical reflectors, made of copper, polished and silvered within their concavity. They are set in two rows, four above five, arranged in the arc of a large circle. A lamp is placed in the focal centre of each, the smoke from which is led off by a tube, passing through each reflector to a common chimney behind.

This lower light is chiefly of use to ships when near the island. As long as it continues in sight, when approaching the shore, they are safe ; but the moment it is shut in by the intermediate summit of the precipice, they are in dangerous proximity to the rocks, and must haul off till they see it again.

The fogs, which are so prevalent on this coast in winter, are the most fatal occasions of shipwreck.

It is then in vain that the watchful keeper trims the lamp, and in vain the inventions of optical science are employed to magnify the light. The dense and blinding mist absorbs the rays, and intercepts the friendly warning. About three years ago, the keeper informed us, a vessel came ashore in a dense fog on the rocks just below the lighthouse. All the crew took to their boat, but were never afterwards heard of, being doubtless swallowed up in the tremendous surf that dashes-in during heavy weather among those rugged rocks. One person alone was saved, a sailor-boy, but a passenger on board this craft. The boat had put off without him; but the crew, on discovering that he was left behind, told him to jump overboard, and they would pick him up. He, however, was afraid to do this, as he could not swim; preferring to take his chance where he was.

The poor lad remained on the wreck till morning dawned; meanwhile, the tide had receded, and had left the vessel high and dry upon the shore. He found he could with ease jump down from the bows upon the rocks below; whence, with no great difficulty, he clambered up the precipice, told his sad tale, and met with hospitality and sympathy.

After drinking in the wide-spread prospect lying in a vast circle around, looking by turns upon the long range of English and Welsh coast, upon the sea, sleeping and sparkling in the sun's bright rays, and upon the island beneath, whose whole outline the eye could here take in, almost as if it had been laid down in a map, we cast "one longing lingering look behind," with a moral certainty that we should see that sight no more, and bade farewell to the lighthouse.

It proved indeed a farewell to the little isle itself; for, as we descended, we saw a skiff even now approaching the shore, sent expressly from Ilfracombe to fetch one of our party to a near relative in urgent sickness. There were several points of interest which we had only imperfectly, or not at all, examined ; and we would willingly have spent another day on the pleasant little spot. But this was now out of the question ; the case was pressing, the wind was fair, the boat was waiting at the beach ; we took a hasty leave of our kind and courteous friend, and were in a few minutes skimming the waves, and looking back to the fast-receding rock, where we had spent a few days of almost unmingled gratification.

A Ramble to Brandy Cove.

A RAMBLE TO BRANDY COVE.

THE BATHING-POOL, ILFRACOMBE.

ILFRACOMBE is not a woody place, certainly; trees are
the only element lacking to make its scenery perfect:
yet, a little fresh water, perhaps, a calm shaded river,
or a lake, in the parts where the downs shut out the
sea, might be also wished for; this, with a little more
timber to take off the nakedness of the aforesaid
downs, would be perfection. And yet, at the end of

May, one can scarcely admit, when one looks abroad, that it does lack anything. These rounded hills, covered as they are with turf, so smooth, and of such a tender green, are beautiful in their broad slopes and convexities; and the differences of light and shadow, and of atmospheric tint, as the sun's rays fall varyingly upon them, and as they are now relieved against the cloud-mottled azure of the sky, now recede behind a prominent mass, with a curving valley between, effectually preclude anything like a wearisome uniformity. Then there are the thickets of furze, sitting like dark crowns upon their summits, and groves of young oak and ash here and there in the bottoms, now arrayed in the freshness of new clothes—a livery of a richer and deeper hue than that of the grass, though yet of a tint which has a lively brightness peculiar to a week or two at this season; and the fruit-trees of the orchards, whose blossoming glories have just yielded to full foliage; and the luxuriant vegetation of the gardens; the young peas and beans and potatoes; and above all, the hedgerows: all this gives such a variety of tint, that one forgets in the fulness of admiration that there is but one colour displaying many shades. Surely there is no other colour that could so charm the eye as *green;* none that could bear to be spread over almost all nature; none that we could look upon so continually, in all sorts of shades, not only without weariness, but with ever-new delight and refreshment. The conventionalism of art puts a sort of *taboo* on the tint in painting; but in real nature the eye never grudges the lavish profusion with which it covers the landscape.

The scarcity of wood here makes doubly valuable whatever approaches to grove or coppice we find, especially in our searches for wild flowers. There is a nice little coppice about a quarter of a mile behind the church, on the road to Lee. You enter through a gate, and the wood is over your head; for it covers the steep side of a sharp ridge, that runs upward, till its extremity becomes one of those lofty peaks called tors, that overlook the sea, down perpendicular precipices of rugged rock. The coppice is very little, but it is a pleasant retreat to pass into its shadow out of the dusty road; for as the slope faces the north, no sun falls on it, at least during walking hours. All through May the lower parts have been covered with primroses and hyacinths and dog-violets, which are now yielding to other candidates for our admiration: the red campion and the herb Robert and the dog's mercury are still abundant there. The mossy banks produce two kinds of orchis, the early purple and the spotted palmate, the latter the less common. The sweet and modest flower of the wood-sorrel peeps from under the shadow of the shrubs, and mingles with the beautiful little yellow pimpernel, or loose-strife—a pretty name for a flower. Higher up among the bushes, all tangled with formidable brambles, grows the bilberry, with its delicate, rosy, urn-shaped flowers, and that curious plant the woodrush; and we have found one single specimen in blossom of

a rather uncommon and fine flower, the bastard
balm.

How much it adds to the pleasure of a walk to
have something to search for, no matter whether it
be insects or flowers, beetles or bee-orchises !—the
having an object of desire, the constant hope of finding a prize, you know not what; and now and then
the delight at finding some unexpected, unthought of,
but not unwished for treasure; greatly enhance the
gratification, and associate indelibly agreeable places
with agreeable emotions. Forgive me if I am tedi-
ously garrulous, but I have always loved to cherish
such associations. I can look back for years, and say
with complacent memory, " It was in such a lane, on
such a day, in company with such and such beloved
companions, that I first found this or that rare insect;
it was under such and such circumstances, impressed
upon my recollection with a vividness that can never
be effaced, that I heard for the first time the voice of
a particular bird."

By the way, about that woodrush, a reminiscence
comes over me, when I see it, more amusing than
flattering. The fine, rather imposing appearance of
its broad leaves, as they come up in hollow tufts, enables " 'cute fellows," in the guise of rustics (more
'cute than conscientious), to palm off roots to gardening Londoners, as those of fine bulbous flowers. I
have seen the plant so often in suburban gardens and
areas, cherished up from month to month, and even
from year to year, until patience becomes exhausted,

and we see the collection some fine morning lying in the horse-road, that I am persuaded it is a common trick, and that many a dishonest sixpence is turned in this way. A good many years ago, a fellow in a smockfrock came to my garden-gate in May, when I was tidying up the beds, and offered me a couple of roots of something of which he said he did not know the name, " but it bore a beautiful pink flower at Christmas." I did not know the plant; I was young in gardening (it was a good many years ago), and rather admired the look of the leaf-tufts. I strongly doubted his story all the time, but he was profuse in his praises of the flower, and told me I should have them for sixpence; and so I invested that amount of capital in the enterprise. The fellow looked about, and said, " He could see that master was a good gardener." My old mother, who thought no great things of my talent in that line, averred that the man's flattery had something to do with the purchase; but I declare to you that it had not. However, I watched the plants till Christmas had passed; but no sign of shoot or flower-stalk was sent up; and one morning they disappeared, not to lie in the horse-road, but quietly put away, decently buried beneath the ash-heap, without any dirge.

How you would admire the little pictures that present themselves at every turn ! Here is a scene, as I sit on this mossy bank, and look over the hedge before me. This nearest field, with the wheat in blade, and that next to it, in grass, and almost ready

H

for the mower; the wind sweeps over both, and we trace its course by the eye; but the effect on one is very different from that on the other. On the young corn the waves give a bluish green, a sort of hoary glaucous tint, as they pass, and have not the fairy lightness of the waving grass in flower, over whose gray and russet surface silvery flittings sweep so lightly, that you might imagine Queen Mab and her airy troop were speeding over it. The fields are sloping away in all sorts of ways. I am sure there is a brook down there in that dark corner between the wheat and the grass; I cannot see any sparkling of water, but I know it by the look of the trees; they are so dense, and there is such an obscurity, a blackness, in and under their screening foliage, as only the vicinity of water gives; it is just such a little patch of deepest shadow in the sunny scene as an artistic eye would delight in. There is the dusty, drab-hued road, winding up between those hedges, half-hidden as it winds; the farm-buildings yonder in the bottom; the old church peeping over the hill, the ridges of its triple roof just in sight, and its square gray tower, with a vane at each of its four corners, all pointing the same way; an emblem of what the church ought to be, rather than what she is, with all her ministries directing the soul to Him who is the living way, the only way, to God. The summits of those broken hills close the view as with a wall; but between them there is just a peep of the ever-lovely sea; and a minute vessel far off, making her way up the channel, reminds our human sympathies that our fellow-man, with all his hopes, and fears, and cares, and toils, is there. The winds are sighing round us, and whispering in these quivering hazel-leaves; and many voices are behind us in the copse—sweet voices of sweet birds! How richly mellow the low notes of that blackbird, who has been pouring forth a broken melody for the last ten minutes, as if unconscious that any one heard and admired! and here, close at my elbow, a tiny rogue of a wren perches himself on a twig, and with tail more than erect delivers himself of such a rapid effusion, that one can scarcely help laughing. There is the sweet call of the cuckoo again —cuckoo! cuckoo! How I love to hear that voice; I stand still to listen, and drink in the notes, as if they were the very quintessence of summer.

That principle of curiosity that prompts one always to penetrate as far as possible, and to see all that may be seen, won't let me sit here enjoying this quiet scene, lovely as it is. I must needs climb those heights, and see what that elevation reveals. The little lane (I ought to have said that a bank and hedge, bounding the foot of the shrubby hill-side, make its bottom a lane) presently opens into a pasture-field, steep enough for a pugilist's fasting walk. The edge of the copse that bounds it is blue with the thick spikes of bugle; and here at the lower parts whole patches are radiant with the pimpernel. Except the corn poppy, this is said to be the only scarlet flower we have; and, in truth, it is a little gem, with its

dark-purple centre and bright-yellow pip in the eye.
Three most sweet little flowers,—sweet, I mean, to the
eye ; for, alas ! neither of them is endowed with fragrance,—the pimpernel, the loosestrife, and the ger-
mander-speedwell,—scarlet, yellow, and blue,—grow
in profusion within a few yards of each other here.
I was tempted to try how a bouquet would look composed exclusively of these three. The effect was most
charming, the loosestrife supplying the foliage, of
which the others were lacking. What a pity that
such a garland should be scentless !

As the ground rises, a commanding view of the
town is opened, which, to be sure, offers nothing either
attractive or imposing. What the stranger is most
struck with, is the gray hue of all the houses ; they
look exactly like the dwellings in the New England
States and the British Colonies. The peculiarity is,
that the walls are faced with slates of a silvery-gray
appearance, which bears the closest resemblance to the
clap-boarding with which it is customary to cover
houses in those timber countries.

Higher still, through another field, where the tiny,
yellow heads of the medic are abundant, and that
curious species of potentilla, well called silver-weed,
that looks as if children had been snipping fancy
leaves with scissors out of a piece of French-gray
satin. Before we leave this field, however, let us turn
and look. Is it not a glorious prospect ? Where but
in England could we see such scenery ?

What a sweet calm reigns over all ! It looks like

a land that had never seen strife. Indeed, I think that the peculiar, indescribable air of security that belongs to an English landscape may be in part owing to the happy circumstance, that for centuries her soil has not known the horrid devastations of war. She may be described in the graphic terms which the prophet uses to express the peaceful fearlessness of the land of Israel in the day when the haughty Gog shall come up against it,—a " land of unwalled villages ;" " them that are at rest, that dwell confidently [*marg.*], all of them dwelling without walls, and having neither bars nor gates" (Ezek. xxxviii. 11).

The sight of yonder homestead, peeping from among its surrounding trees of deepest, massiest foliage, put this into my mind. The white buildings just indicated, rather than shewn, nestling in their bower of verdure, have such an air of peacefulness! Well, that fills the bottom of the valley. Then before us and around is a wide amphitheatre of country, chequered with fields of all shapes, of all shades of green, from the dark corn just up, to the emerald hue of the young grass, or the yellow flush of thousands of newly-opened buttercups; and of all shades of brown, some rich and red, where just ploughed, others grayish with the sprinkling of lime that has been cast over them. The column of white smoke that curls upward from the corner of yonder ploughed field, and falls obliquely away in transparent haze, where the peasant is burning the noxious couch-grass, adds to the quiet dreaminess of the scene. The projecting tors on the left hide the town and the sea; but the summit of Hillsborough's great mass is visible,—a noble object; and in the hollow and on the slopes are many pretty white villas with their gardens and pleasure-grounds.

Another push upward, and all this is shut out; and here we are on the top of the naked, rounded down, with the expanse of the Bristol Channel before us, and the sound of its waves surging among the rugged rocks far beneath our feet. Here is the short, close turf, and the pretty scarlet-tipped bird's-foot trefoil, and the rosy, dwarf red-rattle, and the delicately-formed milkwort, all tiny plants that hardly overtop the turf, close as it is. The milkwort is of the blue variety, the deepest, richest ultramarine,—surely by far the most beautiful phase of this varying little flower.

When I got up as high as this, two or three little things of interest occurred to me. One was the finding of a thrush's chopping-block. You are perhaps aware that the birds of this family feed largely on snails, and that they are said to carry their prey to some selected stone, against which they hammer and bang it till the fracture of the shell enables them to pick out the morsel. I never before had personal testimony of the habit, but here was evidence indubitable. Around a stone about as big as my head, and partially imbedded in the earth, were scattered the fragments of perhaps ten or a dozen snail-shells, all of the same species, the pretty banded wood-snail (*Helix nemoralis*); and the smeared stone made it clear

enough how they had been broken. Two or three of
the shells were unbroken ; they had evidently resisted
all the batterings of the bird, and, as a last resource,
he had endeavoured to get as much as he could from
the natural aperture, for the poor snail, in each case,
was cut and nibbled as far as a bird's beak could
reach it.

But these were not the only shells that I found here.
Scattered about on the downs, three or four hundred
feet above the sea, I found several shells of the common limpets that congregate in thousands on the rock
below. Who could have brought these hither ? I
should incline to reply, the jackdaws that I see perch-
ing on the ledges of the precipice yonder. If human
hands had brought them hither, the mollusk would
have been detached quite clean ; and that whether it
were done for the sake of the shell,—as by children to
play with (a most unlikely supposition, however), or
for the sake of eating the animal. In the former case,
the human fingers remove the flesh clean out, and in
the latter case,—*i. e.*, in being cooked,—it drops spon-
taneously ; but in each of these shells the fragments
of flesh remained adhering all round the concavity,
having evidently been picked out piecemeal by an
industrious bird.

What those birds were there is little room left for
doubt, when one remembers the appetite of the *Cor
vidæ* for mollusks, and observes how numerous the
jackdaws are hereabouts. Indeed, it is by far the
commonest bird of the crow kind in the neighbour-

hood. How they manage to get the fast-sticking limpet from his rocky base, however, I am at a loss to imagine. It requires considerable adroitness in a human practitioner to effect the removal with the aid of the point of a pocket-knife thrust under the margin of the shell. And it must be done in a moment, too; for if you give the shell-fish the least warning, he screws down his shell so tight, and brings the force of his adhering muscles to bear so powerfully, that he defies your operations. But the instinct and cunning of all the crows are very great, especially when whetted by hunger.

Portions of the land are enclosed and cultivated at the top of these downs, and a man was ploughing here as I passed. The walls which bound the field, running along to the edge of the cliff, are built of loose, dry stones, in the country manner, affording in the crevices root-space for many wall-loving plants. The pretty pale yellow heads of the clover-like flower, called lady's fingers, was growing on them in profusion, as it does in the clefts of the rocks all about these precipitous shores, embellishing their ruggedness with its delicate blossoms. Out of the side of the stone wall also grew numbers of tall and noble foxgloves—that most magnificent of British flowers: several of them were already in blossom, though it was yet May; and indeed that reminds me that most of the flowers here appear earlier than the earliest period assigned to them in Hooker's " British Flora." Surely a finer spectacle than a group of foxgloves, growing robust and in full bloom, with the rich purple of their deep cups—the fingers of the folks' (or fairies') gloves—deepened by full exposure to the sun's light, one can scarcely wish to see in the way of flowers. A dozen or so of spikes, all straight and well grown, and tied together in one or two places, and set in a deep vase, make a noble bouquet for the drawing-room. The buds progressively unfold; and as they enlarge they push against one another, and fit themselves into the intermediate spaces, so as soon to make a compact mass, as if all were growing from one stalk; and a truly grand affair it is.

Have you ever seen the *Machilis maritima?** Per-haps you have not, for it is not at all a common insect, and is found only in certain localities, as upon rocks and stones by the sea. It is a rather curious creature, and worth a moment's examination, if you ever fall in with it. I found it the other day near Watermouth, when ascending the face of the cliff, by means of holes, which I had to cut in the soft slate with my knife as I went up; but my situation then prevented me from attending to it. Here, on the stones of this wall, I observed several more leaping nimbly about, and one of these I caught, despite his agility. From their shape, and especially from their being clothed with shining, silvery scales, the insects of this family have

* I must crave the kind tolerance of my readers for occasional repetitions. This *Machilis*, for instance, has been mentioned in a previous page. The present volume is a collection of Essays ori-ginally unconnected.

been called fish-insects; and we name this the Many-
footed Fish. It is not particularly small, being nearly
two inches in full length, from the tip of its antennæ
to the extremity of its long bristles behind. The way
in which it performs its vigorous leaps does not at
first appear; but if you look carefully you will find
all along each edge of the belly a row of short, stiff
points, directed forwards, which move on a joint, and
ordinarily lie close to the body. When, however, the
insect wishes to leap, all these little bristling points
are forcibly thrown out at right angles, as if with a
spring-movement; and by the impulse the insect is
projected forward through the air to the distance of a
foot or more. The tail is furnished with several long
bristles, which have been supposed to be the organs
employed in propulsion; but I think this is quite a
mistake. It is a rather pretty creature, marked in
chequers of light and dark gray, and often reflecting
prismatic tints.

From the highest part of the downs a broad slope
of turf, dangerously steep, descends to an abrupt edge,
whence the rock is absolutely perpendicular for three
hundred feet or so, down to a little cove, fringed with
many a pinnacle and projecting ledge, washed by the
sea at high water. Just as I appeared over the summit of the down, a sharp, querulous cry was uttered,
and presently repeated by a number of similar voices,
and up sprang into the air above the cove about a
hundred and fifty gulls. They had evidently been
disturbed from their resting-places—probably their

breeding-places—in the precipice; for on the shelves
and projecting points many more were still sitting,
which, as I approached nearer and nearer the edge of
the cliff, spread one after another their long pinions,
and leaped up on the wing, to add their cries to those
of their fellows above. From their size and colours,
I suspect them to have been of the species known as
the kittiwake, mostly in adult plumage, though with
not a few yearling birds in the company. They soared
round and round, and in and out amongst each other,
calling pertinaciously their three or four sharp notes,
which resembled the cries of young puppies; and now
and then a low, quiet ha! ha! ha! startled the ear,
like the hollow laugh of a person in an under tone,
but close to you. The flight, though flagging, was
powerful; sometimes one would swoop down upon
another, when the assaulted one would shoot away
from the attack with redoubled speed. One in particular manifested much inclination to strike me, if
he had dared: he made many feints of attack, and
evidently wished to intimate that matters should become serious if I did not desist from peeping over the
cliff. I was quite sure it was one and the same bird
each time, for I followed him with my eye through
all his tortuous course, as he sailed away among his
fellows and returned to the assault. He would descend to my level while yet at some distance, and
then would come speeding on in a straight line for
my face, rising just in time to go over my head. Of
course I knew he would not actually strike me, and

therefore did not flinch, though the impetus with which he came would doubtless have made a blow from his sharp strong beak fatal. In all probability his domestic economy, his mate on her weedy nest of three eggs, was placed on a ledge not far from the spot where I was standing. After a while the birds began to disappear, some behind the promontories that bounded the view, some on the sides of the more distant headlands, and some about the cliffy walls of the cove itself, yet ready to take to the air again on the least alarm.

I wanted much to get down to the beach, but searched in vain for any accessible mode. The only means that seemed possible was a fissure, down which a little stream dribbled. It was in many places only just wide enough to allow me to squeeze through, was very rough and full of slaty *débris*, treacherously slippery to the feet. There was, however, a good deal of long grass, and I thought I could not fall far at a time if I missed my footing, and so attempted it. It was laborious enough; but by sliding in some parts, working with knees and elbows, chimney-sweep fashion, in others, cutting away the soft slate with my clasp-knife, and similar manœuvres, I managed to reach within eight or ten feet of the bottom. There it be-came quite perpendicular: I could easily slip down, but how get up again? There was no other way out of the cove, and the tide was coming in. I was reluctant to lose my labour; and besides wished to see if the exposed rocks would afford me any new zoophytes.

It is true they did not differ in appearance from the rocks of other parts more accessible; but the mind is often apt to conclude that what is difficult to reach must be better than what is obtained with facility. However, it was too hazardous. I waited long before I began to retrace my steps; but at length had to make my way up again through my wet and narrow chimney, dirty enough, and scratched too, when I reached the top.

The man that was ploughing at some distance told me that there was a narrow winding path down the face of the cliff, on the outside of a promontory that he pointed out, and accordingly I went to seek it. I found it a great mass of land, like the main, but almost isolated, being connected only by a narrow pathway, where the cliff on each side comes up to a razor-like edge, about a foot wide. Of course it was grand to look down on both sides at once, from such a giddy height; but indeed, to tell the truth, I tried to think of that as little as possible; and looked uncommonly close to my footsteps till I was fairly landed on the grassy peninsula, where, however, I vainly searched for any trace of a path down to the shore on either side; and therefore had to give it up, *nolens volens*.

And now I dare say, dear reader, you are disappointed that this long ramble of mine led to nothing; and so am I, too; but you see plainly it was not my fault. I strove hard to get down to the tidepools on those rocks that I saw so tantalisingly far below; whence, if I could have got down, I dare say I should have brought up some animals, with whose curious organisation or interesting habits I might have amused you. But let us hope that, since Brandy Cove is so very hard to get at, it does not contain anything particularly worth seeing.

The Sea.

THE SEA.

H.M.S. CYCLOPS SOUNDING IN THE MID-ATLANTIC.

" . . . THIS great and wide sea, wherein are things
creeping innumerable, both small and great beasts.
There go the ships; there is that Leviathan, whom
Thou hast made to play therein."

Oh! it is a glorious subject that mighty sea! When
we stand alone on some lofty cliff, some bold headland
that juts out into the waste of water which roars and
boils in hoarse rage far below, and gaze out to the

I

vanishing horizon. on three sides, with no land to
break the continuity but the narrow strip beneath our
feet, that fades to a blue line behind, an awful sense
of its grandeur steals over the mind. But still more
is this impression heightened to him who, in the
midst of the Atlantic, climbs to the main-topmast
cross-trees of some goodly ship at daybreak, and
watches the bursting of the sun from out of the
sparkling waves. A sense of majestic loneliness in
the vast unbroken waste is felt: the deck is so far
below that it is reduced to a small area, and its
sounds scarcely reach so high; the horizon is immensely expanded; perhaps the winds are hushed,
and the boundless waste is sleeping in glittering stillness; not a speck interrupts the glorious circle: a
solemn awe pervades the devout gazer's mind, as he
recalls the words, " This great and wide sea ! "

I have sometimes pleased my fancy, as I have stood
on the beach of one of our south-western bays, with
the thought, that, if we could send forth a little bird,
with the power of unflagging flight, straight out to
seaward, strictly forbidding the pinion to be closed
until land was beneath her, we might welcome her
again to England, without her course of twenty-five
thousand miles having deviated sensibly from her
original departure. Right away would she stretch,
on something like a S.$\frac{1}{4}$W. course, keeping between
the meridians of 10° and 30° W., across the line on
20°, away through the South Atlantic, crossing the
horrid pole, and then up, up, through the Pacific,

leaving New Zealand on the right and Australia on
the left, over that coral sea, where the isles, though
they look thickly studded on our maps, are widely
enough separated by vast horizons, over the still
more desert North Pacific, in the meridian of 170°
W., across the scattered Aleutian chain, through
Behring's Strait, and over the Arctic pole, giving
as wide a berth to Spitzbergen on the one hand, as to
Iceland on the other, till she folded her wings on
our own fair land once more, having performed her
weary stretch of ocean almost in a straight line.

But even this uninterrupted length, vast as it is,
will give us but an inadequate notion of the world of
waters, unless we consider its area also. By what
comparisons shall we grasp an idea of this? It will
take a diligent traveller several years of almost constant railway journeying, to form a tolerably adequate
notion of the extent of England. Then let him essay
to cover the expanse of ocean with Englands; and he
will have to lay down two thousand five hundred, side
by side, and end to end, before the watery plain is
covered. Or let a vigorous pedestrian set out on a
journey to follow the windings of the coast-line,
whithersoever its indentations may lead him: he
may omit the shores of the smaller islands; and yet
a quarter of a century will have elapsed before he
have finished his task, allowing him fifteen miles
every day.

But the "depths of the sea!" What is in that
quiet bosom, that placid, unfathomable heart, far

below the superficial rufflings of the storm ? I have
often looked down from the taffrail of a ship becalmed
in the midst of the ocean,—down, down, into the
clear, pellucid blue,—and wondered how far it was to
the solid bottom, and what sort of a floor it was, and
what was going on in those solitudes. The world
beneath the waters has beauties of its own ; and not a
few observers have remarked the high gratification
with which they have gazed into its recesses, when
these have not been so profound as to be beyond the
exploring power of the eye. In the quiet lagoons of
the coral isles of the South Sea, as a canoe glides over
the smooth surface, scarcely dimpling it with its progression, so transparent is the water that every feature
of the bottom, though many fathoms deep, is distinctly
traced. The groves of living coral, branching in fantastic imitation of the shrubs and trees of the land,
and bearing in their thousands of expanded polypes,
crimson, green, orange, and yellow, what seem to be
brilliant composite flowers in profusion, form a strange
submarine shrubbery of the gayest colours. The gorgeous shells, those fine cones, and cowries, and olives,
that form the pride of many a European cabinet,
are crawling idly over the brainstones and madrepores;
each partially covered with its fleshy mantle, and
expanding its broad undulating foot, which are glittering in still richer painting than even the porcelain
shells. Long ribbon-fishes, that gleam like burnished
silver, dart by ; and parrot-fishes, coloured with the
bright hues of the birds whose names they bear,

peacefully browse and nibble the young tips of the growing coral. Fantastically-formed little shrimp-like beings, almost as transparent as the water itself, and invisible but for the crimson and violet marks that bedeck their bodies, are sailing or shooting through the weedy groves; and tiny squadrons of pellucid jelly-fishes, and innumerable other strange creatures, now reflect the beam of the vertical sun, and flash into radiance, then relapse into invisibility and secrecy again. Then, like the demon of the paradise, comes stealing along the grim and hateful shark, turning up his little green eye of concentrated malignity, as he passes under your boat, and making your very soul shudder at that gaze.

So again, in the Caribbean Sea, whose crystalline clearness attracted the admiring notice of Columbus, I have stood with delight on the bowsprit of a ship, as she thridded her perilous way through a channel of the coral reef, so narrow as scarcely to allow her sides to pass without rubbing, and marked the sea-life that studded those stony walls. Then, emerging upon a deep bay, where the distant bottom of yellow sand seemed only a few yards beneath the eye, I saw the dark-purple, long-spined *Echini*, and vast, slugglish, red *Urasters*, and huge *Strombi* and *Cassides*, go straggling along; while here and there some enormous tree of coral, or shapeless mass of brown sponge, rose from the sandy waste, like solitary bushes in the desert, and flexible corallines waved their long arms to and fro, in the gentle swell of the ocean.

The Sicilian seas, according to Quatrefages, from their habitual stillness and transparency, afford peculiar facilities for exploring the submarine world. As he leans over the side of his boat, the philosopher glides over plains, dales, and hillocks, which—in some places naked, and in others carpeted with green or brownish shrubbery—remind him of the prospects of the shore. The eye distinguishes the smallest inequalities of the piled-up rocks, plunges a hundred feet deep into their cavernous recesses, and clearly discerns the undulations of the sand, the worm-holes of the rugged stone, and the feathery tufts of seaweed, defining all with a sharpness that seems to reduce to nothing the intervening stratum of fluid, and makes the observer forget the unearthly character of his picture. He seems to be hanging in mid-space, or looking down, like a bird from the air, upon the landscape below. Strangely-formed animals people these submarine regions, and give animation to them. Fishes, sometimes singly, like the sparrows of our streets, or the warblers of our hedges, sometimes uniting in flocks like starlings or pigeons, roam among the crags, wander through the thickets of the *algæ*, or disperse and shoot away in all directions, as the shadow of the boat passes over them. *Caryophylliæ*, *Gorgoniæ*, Sea-anemones, and thousands of other zoophytes, with flower-like petals, blossom beneath the tempered rays of .the sun, enjoying his undimmed brightness, without his raging heat. The long and feathery kinds stream out from the hollows of the

rock in a homely gray garb by day, but all lustrous
with sparks of living flame by night. Enormous dark-
blue *Holothuriæ* creep slowly along on the bottom, or
mount the perpendicular rocks by means of their
thousand vesicular feet; and crimson and purple star-
fishes stretch out their long radiating arms, or curl
them hither and thither, as they sit on the projecting
angles.

The *Mollusca*, some encased in stony shells, others
whose unprotected nakedness is compensated by their
gorgeous colours or elegant forms, go gliding along;
while awkward, long-legged sea-spiders run over them
in their oblique courses, or pinch them with their far-
reaching claws. Other shapes, resembling our lobsters
and prawns, gambol among the weeds, seek for an
instant the surface, to touch the thin air, and then,
with one mighty stroke of their broad tail-plates,
instantly disappear, with the rapidity of birds, under
some friendly arch or overhanging tuft. And strange
beings are there, unknown to our colder seas : the
Salpæ, curious mollusks, of glossy transparency, which,
linked together, form long swimming chains; *Beroes*,
like globes of pure crystal, marked with meridian
lines; *Diphyes*, so transparent as only to be distinguished from the water in which they float when the
eye catches the reflection of light from their sides;
and *Stephanomiæ*, long wreaths or strings of glassy
flowers, adorned with bright tints, but so evanescent
that, when transferred to a vase, they presently wither
away, and leave no trace, no cloud, no sediment be-

hind, to tell that a living form had recently tenanted that vacuity of clear water.

Not as on the land, where the charm of variety is chiefly given to the landscape by the vegetation, the luxuriant apparel of the submarine prospect is mainly dependent on the profusion, the gaiety, and the elegance of the animal life; and this particularly in the warmer seas. Characteristic as is the luxuriant development of vegetable life of the sea-floor in the temperate zones, the fulness and multiplicity of the marine *Fauna* is just as prominent in the intertropical and subtropical regions. Whatever is beautiful, wondrous, or strange in the great and populous tribes of fishes mollusks, crustaceans, stars, jellies, and polypes, is crowded into the tepid and glowing seas of the tropics, rests on the smooth white sands, clothes the rough cliffs, clings, even when the space is before occupied, parasitically to the tenants already in occupation, or swims through the free depths and warm shallows ; while the vegetation holds a very subordinate rank, both as to variety of form and species, and also as to abundance of individuals. It has been recognised as a law in the upper world, that animal life, being better adapted to accommodate itself to outward circumstances, is more universally diffused than vegetable life, or at least can survive the privation of conditions ordinarily essential to vitality, longer than vegetation ; and hence we find the sub-polar seas swarming with whales, seals, birds, fishes, and immense multitudes of invertebrate animals, when every trace of vegetation has disappeared in the rigorous climate, and the frosty sea nourishes no sea-weed in its bosom. The same law appears to prevail in the depths of the ocean; for, as we descend into its profound recesses, vegetable life ceases at a moderate depth; while from the recesses to which no ray of light has ever struggled, *Foraminifera, Infusoria,* and other classes of animal existences, are brought up by the sounding-line in vast profusion.

Sir Arthur de Capell Broke has drawn an interesting picture of the singularly transparent sea on the coast of Norway. "As we passed slowly," he observes, "over the surface, the bottom, which here was in general a white sand, was clearly visible, with its minutest objects, where the depth was from twenty to twenty-five fathoms. During the whole course of the tour I made, nothing appeared to me so extraordinary as the inmost recesses of the deep, unveiled to the eye. The surface of the ocean was unruffled by the slightest breeze, and the gentle splashing of the oars scarcely disturbed it. Hanging over the gunwale of the boat, with wonder and delight I gazed on the slowly moving scene below. Where the bottom was sandy, the different kinds of *Asterias, Echinus,* and even the smallest shells, appeared at that great depth conspicuous to the eye; and the water seemed, in some measure, to have the effect of a magnifier, by enlarging the objects like a telescope, and bringing them seemingly nearer. Now, creeping along, we saw, far beneath, the rugged sides of a mountain

rising towards our boat, the base of which, perhaps,
was hidden some miles in the great deep below.
Though moving on a level surface, it seemed almost
as if we were ascending the height under us; and,
when we passed over its summit, which rose in appearance to within a few feet of our boat, and came again
to the descent, which on this side was suddenly perpendicular, and overlooking a watery gulf, as we
pushed gently over the last point of it, it seemed as if
we had thrown ourselves down this precipice; the
illusion, from the crystal clearness of the deep, actually producing a start. Now we came again to a plain,
and passed slowly over the submarine forests and
meadows, which appeared in the expanse below; inhabited, doubtless, by thousands of animals, to which
they afford both food and shelter,—animals unknown
to man; and I could sometimes observe large fishes
of singular shapes gliding softly through the watery
thickets, unconscious of what was moving above them.
As we proceeded, the bottom became no longer visible;
its fairy scenes gradually faded to the view, and were
lost in the dark green depths of the ocean."

But none of these peeps beneath the surface give
us the slightest idea of the depths of the ocean. Where
and what is the ocean floor in "blue water?" Until
within a very few years this question remained without an answer, and deep-sea soundings were only a
delusion and a snare. Many enterprising officers in
the navies of Europe had made essays to get bottom
in the open ocean; some with the common deep sea-

line, some with spun-yarn, and some with a slender
thread of silk; but all had proceeded upon the assump-
tion that, as soon as the weight touched the bottom,
either the shock would be perceptible to the hand, or
the line would instantly slacken, and cease to run off
the reel.

These assumptions were, however, fallacious. It is
found that the diminution of weight, caused by the
resting of the lead, when vast lengths of the line are
out, is not perceptible to the human hand; and, moreover, that there are currents in the profundities of the
sea, which belly-out and carry away the line long after
the plummet is at rest; and this even when, owing to
the freedom from current of the superficial strata, the
line appears to be perpendicular. Thus immense
lengths of line were run out, but no satisfactory
soundings were obtained.

Then other devices were projected. One thought
that a charge of powder, in a sort of shell, might be
exploded by the shock of striking the bottom; and
that the reverberation being heard at the surface, a
judgment might be formed of the depth, from the
rate at which sound is known to travel through water.
But the experiment did not answer expectation. The
shell exploded, but the surface gave no sign. Sound-
ing-plummets were constructed, having a column of
air within them, which would indicate the amount of
pressure to which it had been subjected. In moderate depths, these answered well; but, in great deeps,
just when their aid was wanted, they failed; for the

instrument could not be constructed of sufficient strength to withstand the enormous pressure of a weight equal to some hundred atmospheres.

It was proposed by one mechanician to adapt the principle of the magnetic telegraph to deep-sea soundings. The wire, properly coated, was to be laid up in the sounding-line, and to the plummet was attached machinery, so contrived that at the increase of every hundred fathoms, and by means of the additional pressure, the circuit would be restored, and a message would come up to tell how many hundred fathoms the plummet had travelled down. This brilliant idea could not, however, be made sufficiently simple for practical avail.

Lieutenant Maury had a curious contrivance executed under his own direction. To the lead was attached, upon the principle of the screw-propeller, a small piece of clock-work for registering the number of revolutions made by the little screw during the descent; and it having been ascertained, by experiment in shallow water, that the apparatus in descending would cause the propeller to make one revolution for every fathom of perpendicular descent, hands, provided with the power of self-registration, were attached to a dial, and the instrument was complete. Mr Maury says that it worked beautifully in moderate depths, but failed in blue water, from the difficulty of hauling it up if the line used were small, and of getting it down if it were large. But we do not see, from his description, how it was to be known when the plummet was at the bottom.

As in all such cases, difficulties and disappointments only stimulated invention. Somebody suggested that a quantity of common wrapping twine, marked off into lengths of a hundred fathoms, and rolled on a reel in a definite quantity, would make a good deep-sea line, with a cannon-ball for a plummet. It was thought that as soon as the ball was on the bottom, the reel would stop; then the twine being cut away, and the remainder measured, the length run off would be known, and the depth obtained at the cost of a cannon-ball and a few pounds of shop-twine. The simple suggestion was presently adopted, and some very deep casts were reported; 34,000, 39,000, 46,000 and 50,000 feet of line were run off, but no bottom found, except in the third of these cases, upon which circumstances afterwards threw doubt. It was only now discovered that in great depths the line would never cease to run out of its own accord; so that there was no means of knowing whether the shot had reached the bottom.

These experiments were not, however, lost labour. For by invariably using a ball of the same form and weight, and twine of the same make, it was found that the rate of descent was according to a regularly diminishing scale. This having been well ascertained, it could be determined with approximate accuracy when the shot ceased to carry out the twine, and

when it began to run in obedience to the current
alone; for this latter power was uniform, while the
former was regularly retardating.

Though the depth of the profound sea was thus
ascertainable, no tidings as yet had come up from it.
The ball and twine were sacrificed, as it was imprac-
ticable to weigh the ball with so slight a thread, from
so vast a depth. But a beautiful contrivance was
now invented by Lieutenant Brooke, U.S.N., by which
the long desired object was at length achieved, and
specimens were brought up from the very floor of the
ocean. It is a most simple affair. The ball (a 64lb.
shot) is perforated perpendicularly, to admit a rod
(which is hollow at the end, and armed with grease),
to slide freely through it. The rod at its upper end
bears two arms working on hinges, to which the
sounding-line is attached, and which, while the line
is strained, are kept projecting obliquely upwards.
A tape suspends the ball, fastened by two rings,
which are slipped over the ends of the arms. The
moment the end of the rod touches the bottom, the
line slackens, the arms drop, the rings slip off, and
the ball is loose. Then the rod alone is drawn up,
with a specimen of the sand or mud, or whatever else
may be at the bottom adhering to the " arming," as
the grease is called.

What, then, is the result? That in no case in
which reliable soundings have been obtained does the
depth exceed 25,000 feet, or something less than five
miles. This is in the North Atlantic; but experi-

ments are yet far too few to allow it to be predicated with certainty that much greater depressions do not exist in other oceans.

Across this ocean it is found that a remarkable causeway or elevated ridge of table-ground runs, connecting the shores of the British Isles with Newfoundland. The availability of this causeway for a submarine telegraph was instantly seen, and it has received the name of the Telegraphic Plateau. The bold attempt to connect the two sides of the ocean with an electric wire, its transient success, and its subsequent failure, are fresh in the minds of my readers; and I need not further allude to these facts, except to say that, in the judgment of men best acquainted with the subject, there is no doubt of the practicability of the scheme, when certain elements of failure, already recognised, are eliminated.

According to Maury, the coating of iron wire coiled around the conductor should be omitted, as serving no good purpose; as immensely increasing the size and weight, and therefore the difficulty of manipulation, as well as the cost; and as throwing a needless strain upon the straight conducting line of copper wires. He would adopt the " Rogers cord," which consists of a conducting-wire braided, whipcord fashion, with bobbin or twine, after insulation, and then protected with a cement, which shields the gutta-percha from injury; the whole cord being so slender and easily handled that a single ship may carry the whole, and "pay" it out as she proceeds. The weight of the

Rogers cord is so slight, as to carry it down at the rate of a mile or two per hour ; it is not stouter than the ordinary log-line, so that it can be readily paid out. The amount of " slack" required to feed the currents is not nearly so great as is generally supposed, because the set of the Gulf Stream lies so nearly parallel with the course of the wire, that for a great part of the way the current would scarcely throw the cable out of its proper line. Supposing, however, a current of two knots an hour, for the entire distance, and its course to be at right angles to the cable, the cord, being paid out with ten per cent. of slack, will sink at the rate of two miles an hour : the current may be granted to extend to the maximum depth of half a mile : any given part of the cord, therefore, as it goes out, occupies a quarter of an hour in sinking through this distance. During this interval alone is it subject to the current, which sweeps it half a mile to the left of the ship's course, going eastward ; after which it sinks perpendicularly through the still water, till it reaches the bottom.

The result would be, not a sinuous, but nearly a straight course, only running uniformly half a mile to the left of the track of the ship.

But what proof is there of the existence of such a stratum of still water at the bottom? A beautiful and convincing proof, derived from the organisms that have been brought up from this very plateau by Brooke's sounding apparatus—its first trophies. The naval officer who made the casts, removed from the cup of the rod a little column of what he judged to be a smooth unctuous clay. This, according to his instructions, he carefully labelled and preserved, and on his return to port transmitted the specimens to the proper board. They were immediately sent for examination to eminent microscopists in Europe and in the United States, and proved to be of great interest. The whole of the little packets of supposed *clay* were found to be actually composed of minute shells of microscopic animals, not a particle of sand or gravel or mud being discoverable among them. The great majority of these shells were of a calcareous nature, and belonged to that group of lowly animals known as *Foraminifera*. There were, however, among them a few siliceous shells of those disputed organisms which are so keenly occupying the attention of microscopic savans,—the *Diatomaceæ*. These exquisitely formed shells consist of films of lime and flint, so delicate that a very little abrasion, a very slight degree of violence, is sufficient to break them up into minute fragments; yet the specimens *were almost uniformly perfect*. The inference is then irresistible, that on that quiet floor the countless generations of little shells lie as they fall, gently dropping, like the soft flakes of snow on a calm winter's day, through an atmosphere of water whose density no motion agitates, where there is not current enough to rub their tender forms one against the other, nor to sweep among their millions a grain of the finest sand, or the least atom of gravel from the steep sides of the Grand

K

Bank, that rises like a vast mountain of rock from
the very edge of the plateau.

Professor Bailey, who examined these deposits,
assumes that these countless hosts of animalcules did
not die, much less live, on the spot where they are
found. It is probable that at that vast depth total
darkness reigns perpetually, no ray of light from the
sun having power to struggle through a layer of water
two miles in thickness. Could they bear this privation? It is scarcely supposable that their tender
tissues could sustain the pressure of so great a column
of water, equal to the weight of four hundred atmospheres. In all probability they lived near the surface, perhaps finding their range of motion and their
support in the immense fields of floating weed, the
Sargassum, that cover the area of the Gulf Stream,
—that wondrous mighty river of warm water that
pursues its unerring track through the broad Atlantic,
as steadily, and within as well-defined bounds, as the
Thames through the plains of Middlesex, or the
Amazon through the forests of Brazil. Here, on the
countless stems and leaves and vesicles of the yellow
weed, amidst a vast profusion of other animal life,
they probably sported, enjoying the genial influences
of tropical light and heat, and carrying with them, in
the warm surface-waters of the Gulf, the same favourable conditions of existence, long after the swiftly
speeding stream had carried them beyond the tropical
latitudes.

But, day by day, hour by hour, ten thousand times

ten thousand of the tiny population, populous beyond all parallels drawn from the dense crowds of London or the teeming millions of China, were dying; and as they died, they slowly fell from the floating weed, and, partially sustained a while by the gases formed in their decomposing tissues, during which the superficial currents might softly waft them many a league, they at length reached the distant bottom. Then gently dropping, perhaps on some huge anchor, or water-logged hull, their never-ceasing accumulations would gradually hide the mass under a fleecy covering, "presenting the rounded appearance which is seen over the body of the traveller who has perished in the snow-storm."*

Other specimens have since then been obtained from other seas. From the Gulf of Mexico, the Caribbean Sea, the vicinity of Kamtchatka, Behring's Straits, and the region south-east of Papua, the ocean-bottom has yielded samples. From this last locality, at a depth of thirteen thousand feet, the remains of abundant animalcules come; but they are of a different class from those which occupy the North Atlantic, the calcareous *Foraminifera* being almost wanting. Instead of these there occur the strange shells of *Polycystina*, and some *Diatomaceæ*, but principally the flinty spicula of Sponges. Various forms of these occur, but mostly of the types which we are familiar with in our native species; long straight needles, fine drawn spindles, glass-headed pins, and three-rayed stars.

* Maury.

This result is interesting. These seas are full of
coral-reefs ; they are the very metropolis of the corals
and madrepores. To these is allotted the duty of
separating the lime held in solution by the sea-water,
and to the mollusks, whose massive shells swarm on
every bank, and form a broad white band or long
high-water mark on every beach. These artisans almost monopolise the lime-works of the South Pacific,
and leave comparatively little calcareous matter for
the chambered and perforated dwellings of the tiny
Foraminifera. On the other hand, the flint-glass
workers find a fair field for their delicate chemistry,
and spin their brilliant structures unimpeded. But
there seems less of the siliceous than of the calcareous
element in the warmer seas, and these operations are
there comparatively few.

Here, again, the microscope bears witness to the
perfectly uninjured condition of the majority of these
very fragile organisms. Some of the shells even retained their soft fleshy parts when subjected to examination. It does not follow as absolutely certain, however, that they were alive when collected at such vast
depths. The enormous pressure of the superincumbent water may have a tendency to prevent, or at
least to retard, decomposition ; and the bodies, if
they, in any cases, sink so rapidly as to reach the
great profundities before the soft parts are dissipated,
may possibly retain them for an indefinite period.
However this may be, it is interesting to find the
same testimony to the uninterrupted stillness of the

depths of ocean in these antipodean regions, as was recorded in the northern half of the Atlantic; and especially when, as was the case, results exactly similar were yielded by the casts obtained from the icy seas of Kamtchatka and Behring's Straits. Here, too, the deposits are wholly siliceous, and are principally rich in the remains of the *Diatomaceæ*.

While these results were being obtained with the newly-invented sounding apparatus of Brooke, H.M.S. *Herald* was engaged on a surveying cruise in the Pacific; and her surgeon, Mr Macdonald, an accomplished naturalist, was pursuing similar investigations of the deep-sea bottom. He found the *Foraminifera* in very considerable abundance in the vicinity of the Fiji Islands, at a depth of upwards of six thousand feet; and, what is a fact of great interest in connexion with these vast burial-grounds, he observed considerable numbers of the living animalcules adhering to the fronds of the smaller marine Algæ, either floating on the surface of the ocean, or growing on the shores of the Pacific Islands; so that the abundant appearance of the dead shells of these tiny animals in the sand of every beach, and in every sea-bottom fathomed by the armed lead, was satisfactorily accounted for. How inconceivably numerous these remains of animal life really are in the sands of the shore, may be estimated from the fact, in addition to that already mentioned, that in some beach-sands upwards of half of the entire bulk is composed of these microscopic shells. Plancus counted six thousand in an ounce of sand from the Adriatic, and D'Orbigny estimated the number in a pound of sand from the Caribbean Sea at no less than 3,849,000,—nearly four millions of individual animals !

Mr Macdonald observes, that the spicula of Sponges and Asteroid Polypes, and the minute embryonic shells of *Gastropoda, Pteropoda,* and *Conchifera,* are usually found with the *Foraminifera* in the soundings which he has examined. The pelagic shells, or those which during life rove freely through the sea, descend into the profound recesses after death by their own gravitation; but the others are washed off from every coast and reef; millions of organic and almost indestructible forms thus combining every day and hour to enrich the dark and solitary bed of the ocean, and to smoothen its rugged floor. The muddy bottom of the sea outside the Capes of Port Jackson is nearly altogether composed of such materials, as is that which fringes a considerable portion of the coast of North America, and other vast regions.

A few particulars of the life-history of these atoms, which play a part so important in the physical economy of the earth, may not be unacceptable to my readers. The older conchologists were acquainted with a few shells of microscopic minuteness, some of which closely resembled in form that of the Nautilus, and, like it, were found to be divided into successive chambers. For a long time these tiny forms were considered as Mollusca, and belonging to that highest type of structure. which includes the Nautilus and other Cuttles; instead of taking their rank, as they are now known to do, among the very simplest developments of animal existence. The chambers communicating by several apertures, they were named *Foraminifera;* and that appellation is now found to have a further appropriateness, from the curious fact that their shells, which are formed exclusively of lime, are perforated with minute orifices, often so numerous and approximate as to impart a sieve-like character of the structure.

About a quarter of a century ago, however, M. Dujardin announced the true condition of these little creatures. Their soft parts consist of a homogeneous jelly or glaire, without any distinction of organs, which fills the chambers with its clear transparent colourless pulp, and is endowed with the power of pushing out irregular prolongations of its own substance in every direction, and from every part of its surface. These prolongations take the forms of expanded films of excessive tenuity, or lengthened threads, of a viscid semi-fluid, which coalesce and unite by contact, or are separated and drawn out in so great an irregularity as to show that they are not enclosed in any skin or membrane. The extensions often reach to a length thrice or four times that of the shell, and may be seen and watched in an interesting mannĕr, when the living *Foraminifer* is placed in a drop of water within the glasses of an animalcule-cell of the microscope, and allowed to remain a few hours perfectly undisturbed. We see the *pseudopodia*, as these pro-

jections are called, protruding their tips from various
surface-apertures of the shell, and then gradually—so
gradually that the eye cannot recognise the process—
stretching and expanding their threads and films of
delicate *sarcode*, till in the course of a few hours these
will be found to reach almost from side to side of the
glass cell. The extension is generally in two opposite
directions, corresponding to the long axis of the shell ;
though the branched and variously connected films
often diverge considerably to either side of this line,
giving to the whole a more or less fan-like figure.
This array, so very deliberately put forth, is very
rapidly withdrawn on any disturbance being given to
the little operator ; as when the water in the cell is
agitated by a sudden jar on the table, and especially
by slightly moving or turning the glass cell-cover.

It is manifest, from distinct, though small, changes
of position in the shell, while these elongations are
going on under observation, that it is by means of
the adhesion to extraneous objects, and the consequent
contraction, of the *pseudopodia*, that the animal drags
its shell along a fixed body. It is remarkable, however, that Mr Macdonald finds the *Foraminifera* in
the Pacific, in general, attached to sea-weeds, and
other foreign bodies, by a short, thick footstalk, somewhat resembling that of the *Lepas*, and so precluded
from the possibility of locomotion. With his very
extensive opportunities of observation on the living
forms in the South Sea, he professes to have " never
been able to discover their branched *pseudopodia*, or

the slightest evidence of the crawling movement which they are reputed to exhibit." In those of the European seas, however, these powers have been seen by too many accurate observers, to leave the slightest doubt of the facts. I have myself kept some of the more familiar British forms in aquaria for months, and have seen them crawling every day (especially by night) over all parts of the vessel and its contained sea-weeds. It may be that Mr Macdonald, pursuing his researches on ship-board, was not able to afford his specimens the continuance of absolute stillness, which is essentially indispensable to their activity.

The sustenance of these simple bodies is secured by the enveloping and adhering powers of the *sarcode*. The *pseudopodia* are food-gatherers as well as instruments of locomotion. They explore the vicinity of the animal, feeling about in all directions; any animalcule, or simple plant, more minute than themselves; any stray Diatom, or Desmid, or Alga, or Infusorium, or embryo Mollusk, or Sponge-gemmule; or any particle of decomposing organic matter, touched; is instantly entangled and laid hold of by these viscous hands: the sarcode envelopes and covers it, and then, contracting, draws it into the interior, where it may sometimes be followed by the eye, through the transparency of the shell. There is no mouth, no stomach, no digestive canal; but the homogeneous jelly appears to have the power of assimilating the nutrient juices of the food in any and every part alike; and hence it is of no consequence what part of the surface is brought into contact with the food; it is *there* embraced, and, as one may say, swallowed, and there digested ; so that any part of the simple glairy body may become a temporary mouth or an improvised stomach. Generally, the residuary portion of the food-pellet is slowly pushed out and rejected at the nearest point of the surface; but not always ; for these exuviæ sometimes accumulate in considerable numbers, so as even to choke up a large part of the cavity of the shell.

Nearly two thousand species of these little creatures have been distinguished, and they are doubtless much more numerous than this ; *all* are not microscopic, some of the oceanic species being of the size of a shilling, and a few even as large as a crown-piece. There is great diversity of form in the shells : some are straight or curved rods ; some conical; some have the shape of elegant vases or bottles ; some are orbicular, many discoïd, and the majority spiral. The shell appears to be invariably simple in its first stage, being deposited around a primal nodule of sarcode ; this is the first chamber : buds develop themselves in succession from this, each of which deposits its calcareous chamber : thus successive chambers are formed. If these buddings take place in a right line, the mature shell will be rod-like, or necklace-like ; but if the axis of development incline slightly to one side, a curved rod, or row of beads, will result ; if this inclination be in excess, a spiral growth will be formed, the character of which will be modified by the ratio of increase of the successive chambers, and by their ventricose or parallel-sided form. A very prevalent type in the Pacific is that of the *Orbitulites*, which very much resembles a coin in its circularity, flatness, and comparative thickness; and a species from the Australian coast equals a sixpence in size. This pretty shell is made up of a number of thin concentric circles, each of which is composed of many flattened chambers, communicating by minute orifices with those of its own range, and also of the ranges within and without it. In this type the central or primal cell is comparatively large, of pear-like form, and is almost surrounded by a secondary chamber, which is far larger than any of the rest.

Very closely allied to the *Foraminifera* are the *Polycystina;* shell-bearing animals, of even more extreme minuteness, which have been only recently made known, but which are found to exist, in considerable abundance, in the oceanic deposits, and to be still more numerous in certain geological formations. They have been recognised by Ehrenberg in the chalks and marls of the Mediterranean coasts, as Sicily, Greece, and North Africa; and in the diatomaceous deposits of Bermuda and Virginia: in the island of Barbadoes, the rock of a very extensive district has been found by the great Prussian microscopist to be almost entirely composed of Polycystine shells, with a slight admixture of *Foraminifera* and *Diatomaceæ*, and with calcareous earth, which seems to have been derived from the decomposition of corals;—all oceanic

organisms. Some three hundred species of *Polycystina*
have been detected in the Barbadoes strata, chiefly by
the investigations of Sir R. Schomburgk. The class
differs from the *Foraminifera*, in the circumstance
that the shells are siliceous instead of calcareous;
their forms are even more bizarre, and often possess
remarkable elegance and beauty. A prevailing type
of form is a sort of dome or cupola, with an apical
prolongation of spine, and terminating in three equidistant spines below; their walls beautifully fenes-
trated with large angular or circular perforations,
and, both externally and internally, exquisitely sculptured, so that they have been compared with "the
finest specimens of the hollow ivory balls carved by
the Chinese."

According to Professor Johann Müller, who has
pursued some investigations on the living *Polycystina*
of the Mediterranean, the sarcode is of an olive colour,
which forms pseudopodia, that project through the
fenestral apertures; but which, in a retracted state,
occupies only the upper vault of the dome, and is
regularly divided into four lobes. This is, at least,
the case with some species; but observations on the
trans-European types are still very deficient. These
animals seem almost as widely diffused as the *Foraminifera;* but, from their far greater minuteness, they
have not been so generally recognised.

Important as are the two classes of microscopic
beings of which we have been speaking, from their
vast numbers, and the office assigned to them in

effecting physical changes in the crust of the globe, far more inconceivably numerous are the hosts of the *Diatomaceæ*, and far more momentous are the operations they perform and the influences they exert, both on the world and its inhabitants. To those of my readers—a very considerable class, I doubt not—to whom a *Diatom* is but a Greek compound, I may be permitted briefly to explain the more obvious characters and attributes that distinguish this universally distributed, yet recondite, tribe of organic entities.

By the general, but not quite universal, consent of microscopical science, the *Diatomaceæ* are plants, each composed of a single cell, invested with a coat or shell of pure silex (flint), endowed with spontaneous motion, and mostly found in aggregation of many individuals, so attached in regular series as to form chains, more or less readily separable. The endochrome, or vegetable pulp, which in most plants is of a green hue, is always in this class of a golden-brown or yellow, and its particles have occasionally a sort of circulating movement within the cell.

The shell, or frustule, has a fixed form and dimension in each species, though these are subject to very great diversity in different species. Its shape is often extremely elegant; and its glassy surface is exquisitely sculptured into pittings or prominences, which are arranged in the most elaborately varied and beautiful symmetrical patterns.

Perhaps the most ready mode of conceiving of these creatures, by one who has never viewed them with

the microscope, will be to take a low pill-box of card,
and suppose it to be made of flint glass, delicately
sculptured, and reduced to an invisible minuteness.
Suppose the granules of the yellow endochrome to be
enclosed in this box, surrounding a central mass called
the neucleus, which seems to be the very heart, or soul,
or life-point of the tiny organism. The cover and
bottom disks, called *valves*, are very easily separable
from the hoop that unites them (a parallelism that too
often obtains in their cardboard representative;) and
so these single valves are often found alone.

The name *Diatoma*, which, originally given to a
single genus, has been applied to the whole order, has
reference to the readiness with which the strings or
chains in which most of the forms are aggregated,
may be separated, looking as if divided by a sharp
cut, either partially or wholly. And this depends on
their mode of multiplication. For the law by which
these atoms increase is highly curious. The pill-box-
like frustule becomes deeper by the widening of the
hoop, thus pushing the valves further from each
other ; then across the middle two membranes form,
which, by-and-by, by the deposition of flinty matter,
become glassy valves, corresponding to the two outer
valves ; and then the whole frustule separates between
these two new valves, and form two frustules. The
old hoop (in some cases at least) falls off, or allows the
hoops of the new-made frustules to slip out of it, like
the inner tubes out of a telescope.

Now the separation of the frustules thus made is

not always so complete, but that they may remain adherent to one another by some point of contact; and hence arises a most singular and interesting appearance often presented by these bodies. Let us suppose that the original frustule was of the shape of a brick, (for the pill-box form is only one of many), and that by successive acts of self-division, it has formed itself into a number, say a dozen, of bricks. These of course are laid one on another, forming a pile; but all the individuals adhere to one another by a minute point at one corner, and the matter of adhesion is sufficiently tenacious and sufficiently yielding to allow of the brick-shaped frustules moving freely apart in every point except just the connecting angle. It is not the *same* corner that adheres all up the pile; more frequently *opposite* corners alternate with each other, yet not very regularly; and thus an angularly jointed chain of the little bodies is formed, which is very characteristic. In some species, in which the form is a lengthened oblong, the frustules have the faculty of sliding partially over each other; and thus the chain takes the form of a series of steps, of which the length is much greater than the width or height.

Some of the forms have the frustule seated at the end of a long and slender footstalk,—a thread of spun glass, on whose elastic summit they wave and dance with every movement of the waters. The self-division of the frustules here frequently extends to the stalk; and so we find beautiful little fan-like tufts or shrubs, all educed by this imperfect multiplication. In almost

all other cases the atoms possess the power of spontaneous movement to some extent. Often this takes
place by a series of intermittent jerks, which carry the
Diatom onward in a given direction for a while, and
then suddenly ceasing, yield to similar motions in an
opposite direction, by which the progress made is reversed. In some cases, as in the genus *Bacillaria*,
which we have just compared to a column of bricks
sliding one over the other, this movement of sliding
goes on till the frustules are on the point of separating, which then retrace their course till such a catas-
trophe seems equally imminent in the opposite direction.

It is generally considered that no power of choice,
no real volition is manifested by these motions, which
are asserted to be merely mechanical, and not produced by any motile organs, properly so called. But
Dr G. C. Wallich has recently published some elaborate researches made upon the free-swimming *Diatomaceæ* of the South Atlantic, which lead to a different conclusion. He has shewn that particles of extraneous matter lying in the path of a moving Diatom
are occasionally pushed forward by it, or, if behind,
are taken in tow, and dragged after it; the object in
neither case being in contact with the frustule, but
considerably distant from it. The object in tow
accurately exhibits and repeats every jerk, progression,
or pause of the 'tug;' and at times is even drawn up
to it, it may be, from an oblique position, and is then
either released or carried along with it, adhering to

one of its surfaces. And these and similar phenomena occur simultaneously with several remote and independent particles of matter.

These phenomena (for the details of which I must refer my readers to Dr Wallich's own admirable memoir*) are quite inexplicable on the hypothesis of exosmotic and endosmotic action, to which the motions of the *Diatomaceæ* had been referred by the best previous authorities. To explain them consistently, we are irresistibly led to infer the existence of numerous long prehensile filaments; capable of protrusion, of extension and retraction; of extreme tenuity, yet of extraordinary strength and elasticity; in virtue of which both the ordinary to-and-fro movements, and the secondary motions affecting surrounding bodies, are performed. It is true, no trace of such filaments can be detected with the highest powers of magnification yet brought to bear on them; but the inference of their existence from the phenomena recorded seems unavoidable, or, in other words, the phenomena are inexplicable on any other hypothesis.

To me it appears that the whole of these observations, though they do not settle the point of the systematic rank and position of these organisms, do add a considerable weight to the opinions of those naturalists who refer the *Diatomaceæ* to the animal rather than the vegetable kingdom; the presumed retractile and extensile processes bearing a very distinct analogy

* *Annals and Magazine of Natural History for January*, 1860, p. 15, *et seq.*

L

to the *pseudopodia* of the *Foraminifera*, and to the whip-like filaments of many *Infusoria;* while their parallelism to any organs known to exist in plants is much more vague and remote.

We may leave this question to be settled by others. The decision will not affect the wondrousness of the facts connected with the economy of these almost inappreciably minute beings; namely, that they, far more than any other created beings that we are cognizant of,—incomparably more than the lions and tigers, the bulls and behemoths, the rhinoceroses and elephants, the cachalots and whales, far more than even busy man himself,—are the master-builders, to whose unceasing agency God has committed the task of manufacturing, of augmenting, and of variously modelling this immense κόσμος of our present residence. Inhabiting all waters, and swarming in rivers, estuaries, and lakes to such an extent, that their siliceous shells, by constant deposition, block up old harbours, narrow and spoil navigable channels, and form enormous beds and strata of earth,—it is yet in the high seas that these innumerable artisans have their great workshop; it is in the ocean, boundless and fathomless, that the grandest processes are going on of their stupendous handiwork.

Far up in the frozen north, and where the mighty barrier of eternal ice forbids the approach of man to the antarctic pole, the tiny Diatoms are building their Cyclopean masonry, and laughing to scorn the castings of our mightiest furnaces, and the forgings of our

Nasmyth hammers. Sir James Ross found the surface of the Southern Sea bordering that ice-barrier
thick with a brown scum, which consisted almost
exclusively of living *Diatomaceæ;* and Dr Joseph
Hooker remarked that they were rendered peculiarly
conspicuous by their becoming enclosed in the newly
formed ice, and by being washed up in myriads by the
sea on the pack and bergs, everywhere staining the
white ice and snow with their own ochreous brown
hue. A deposit of mud, consisting mainly of the flint
shells of these beings, extending not less than four
hundred miles in length and a hundred and twenty in
breadth, was found at a depth of from two hundred
to four hundred feet, on the flanks of Victoria Land,
in 78° south latitude. The depth and thickness of
this deposit could not be conjectured; nor do we know
anything of the rate at which it increases; but observations in future ages may determine this from now
known data, and an estimate may then be formed of
the scale on which these laborious operators turn out
their work.

Every frustule of the *Diatomaceæ* adds its quota to
the solid structure of the globe, and that whatever the
destiny of the living being. It is not only those which
die what a jury of Diatoms might call a natural death,
not only those that fall quietly to rest in their bed,—
the mighty quiet bed of the ocean, that are adding
their shells to the globe-crust: those incalculable
millions of millions that form the sustenance of millions of hungry and predatory animals, all come to the

same end at last; for the silex of their frustules is un-alterable and indestructible. And here we obtain a glimpse of the exceeding wonderful economy of crea-tion; we see with adoring admiration how strangely wise and well-arranged are His plans,—the Lord of Hosts, who is wonderful in counsel and excellent in working.

Guano, that potent manure which has so increased our crops, consists, as everybody knows, of the dung of sea-birds. For ages before the discovery of America the careful Peruvians had collected it, and employed it in their fields and gardens. It was guarded by rules of the most rigid economy. Laws, sanctioned by the punishment of death, forbade the killing of the young birds. The guano islands were all enrolled; each was put under the care of a government inspector, and assigned to a certain pro-vince. The whole tract of country between Arica and Chancay, a distance of two hundred and forty miles, was exclusively manured with guano; and to a certain extent these traditionary customs are still maintained in Peruvian agriculture.

To turn to European consumption, we find the results not less important. From one island alone, a stratum of guano, thirty feet in thickness, and cover-ing an area of 220,000 square feet, has been entirely removed within twenty-seven years. In one single year (1854), the enormous amount of 250,000 tons of this accumulated excrement was dug in the Chincha Isles, and the actual annual exportation doubles that quantity. Thus, the dung of wild ocean birds yields a larger revenue to the Peruvian exchequer than all the silver mines of Cerro de Pasco; and its transport occupies greater fleets than ever Spain possessed at the proudest height of her maritime ascendancy.

Now *Diatomaceæ* form a very considerable percentage of the entire bulk of this substance, the value of which is augmented in proportion to the abundance of these microscopic organisms. Great masses may often be found wholly composed of the aggregated frustules of Diatoms. How are these procured in such vast supply? It has been by some supposed that the birds, or that fishes on which they subsist, feed *directly* on them. But this is manifestly untrue, as Dr Wallich has shown, since, with one or two rare exceptions, Diatomaceous frustules are sufficiently large to be appreciable by any bird's eye. Nor could any vertebrate animal we are acquainted with, by any possibility, gather together, within a reasonable period, a sufficient supply of such infinitesimally minute nutriment as these organisms afford, even supposing the optical difficulty to be overcome. Nor could any prehensile or masticatory apparatus deal with it, if taken into the mouth: it must be swallowed *en masse*.

But the intervention of swarming hosts of invertebrate animals solves the difficulty. It is well known that the vast tribes of bivalve *Mollusca* are supported almost wholly on these and similar entities; which are taken, without any craft, or violence, or pursuit, or even selection, by the mere action of ciliary currents bringing the floating organisms to the gaping stomach. There are, moreover, lower forms than these, but of kindred structure and appetites, as the *Tunicate Mollusca*, which devour immense multitudes of microscopic creatures; and these tribes are numerous and varied. Some of these are free rovers in the ocean, as the *Salpadœ*, and these occur in hosts only less wonderful than the Diatoms themselves.

Dr Wallich speaks from his own experience, confirmed, however, by many other observers, when he says, that between the Cape of Good Hope and St. Helena, for *many degrees* of latitude, the ship passed through vast layers of sea water so thronged with the bodies of a species of *Salpa* as to present the consistence of a jelly. These layers extended for several miles in length. Their vertical depth it was impossible to ascertain, owing to the motion of the ship. They appeared, however, to extend deep; and in all probability were of a similar character to the immense aggregations of close-packed swimming invertebrata so well known to mariners in Arctic regions under the appellation of "whale-food." Each of these *Salpœ* measured about half an inch in length; but so close was their accumulation, that of the quantity collected by a sudden plunge of an iron-rimmed towing-net, *half the cubic contents*, after the water had drained off, generally consisted of nothing but one thick gelatinous pulp.

The stomach in these translucent and generally colourless creatures forms a minute, opaque, yellow

ball, which, being opened, is found to be filled with
Foraminifera and *Diatomaceœ*, from which latter it
derives its colour. A very large species of *Salpa*,
measuring some six or seven inches in length, is found
in the equatorial regions of the Atlantic, whose proportionally larger digestive cavities are filled with
Rhizoselenia, a tubular form of Diatom occurring in
vast profusion there. "The alimentary matter of the
Salpœ," observes Mr Macdonald, "is composed of animal and vegetable* elements in nearly equal proportions ; and when the microscope reveals the calcareous
shells of *Foraminifera*, the beautifully sculptured
frustules of *Diatomaceœ*, keen siliceous needles [of
Sponges], and the sharp armature of minute *Crustacea*,
within an intestinal tube so tender and friable that it
withers at the human touch,—one cannot help admiring the operation of those conservative properties with
which its delicate tissues are endowed. Each atom
yields to acute impression as by an instinctive intelligence, evading injurious contact ; and although a con-
tractility of the tube is essential to the due perform-
ance of its functions, no evil thus befals its integrity
till the term of life is at an end."

The digestive action of the Mollusk effects no change
in the earthy constituents of its food ; and thus the
calcareous shells, and the siliceous spicula, and frustules, lie uninjured in its stomach, disjointed and
broken, perhaps, by trituration, and cleaned of all

* That is, assuming the *Diatomaceœ* to be plants, according to
the received doctrine ; but *vide supra*.

soluble matter, till they are ejected in the fæcal pellet, to be dispersed and carried down individually to the still, and silent, and sombre ocean-floor.

When we consider the immensurable multitudes of these molluscous animals that throng the seas, which feed almost exclusively on the organisms I am speaking of, we shall see how immense a quantity of inorganic matter (yet of organic origin) is every moment being discharged into the sea, and every moment arriving at the bottom. But a very large proportion arrives at the same terminus by other stages, considerably modifying its conditions and ultimate form. The *Salpæ* and similar creatures form the main food of millions of voracious fishes. The shells and frustules of lime and flint contained in the stomachs and intestines of the former are received into those of the latter; and, passing this ordeal uninjured, as well as the other, are in like manner discharged, after digestion, free from their own organic contents, and those by which they were enveloped. But these pelagic fishes are preyed upon by pelagic birds; and the Diatoms and Foraminifers pass into the stomachs of these clamorous sea-fowl, and form the basis of the guano which is ever accumulating on the whitening rocks.

Again, these soft-bodied *Mollusca* constitute the principal sustenance of the giant *Cetacea*. The wallowing whale, or the huge cachalot, drives, with expanded jaws, into such a shoal of close-packed *Salpæ*, as Dr Wallich describes; then, closing his enormous mouth, he lazily entombs myriads of the soft unresisting prey, and repeats the action till his vast stomach is full,—a great cauldron of living jelly. The jelly soon disappears under the solvent action of the gastric juice, and becomes the seething blood of the leviathan; but the minute shells and frustules still travel unharmed ; the heat, the maceration, and the acid have no power to dissolve *them*, and they at length come forth from *this* ordeal as safe as from any former one.

But it is probable that these siliceous and calcareous atoms do not pass from the intestinal canal of the *Cetacea* in individual isolation. They are individually unchanged in form and structure, but are in all likelihood aggregated and conglomerated into cohering masses, each mass homogeneous in its kind. Siliceous particles, in particular, are known to have a power of cohesion, with considerable tenacity under certain conditions; among which pressure, and an animal cement, may be adduced. Professor Bailey, of New York, found some masses of siliceous matter, obtained from Diatomaceous deposits, which he in vain endeavoured to break up by boiling in water and in acids, and by repeated freezing and thawing. At length he boiled the lumps in a strong solution of caustic alkali, under which treatment they rapidly split up, and crumbled to a paste composed of the frustules of *Diatomaceæ*.

Let us suppose, now, a school of whales rioting amidst a vast field of *Salpæ*, which, in their turn, have been pasturing on microscopic Diatoms. Beneath them,—

" A thousand fathoms down,—"

lies an ocean-floor of soft cretaceous clay, the produce
of some coral reef, which has been browsed upon and
ground to powder by the molar teeth of myriads of
Parrot-fishes (*Scaridæ* and *Labridæ*) for ages. From
the full-fed whales fæcal pellets are constantly dropping, each of which consists of siliceous matter, resolvable, indeed, into frustules of Diatoms, and shells of
Polycysts, and spicules of Sponges, but now concreted
into an irregularly nodulous, compact mass. These fall
on the soft, calcareous, pasty bed below, and sink into
its impalpable bosom; the white, creamy semifluid
closing over each nodule, and burying it from all disturbance. Geologic periods pass; upheavings of the
crust roll away the sea into other channels, and the
calcareous bed is a thick stratum of chalk,—the white
cliffs of the Albion of the day. The pickaxe and the
spade go to work, and lo! irregular nodules of flint
appear, and *savans* wonder how they came there. The
hammer breaks them open, and the lapidary, with his
lathe, grinds out a thin section, which the microscopist puts under his best powers. He finds that spicules
of Sponges, and valves and fragments of *Diatomaceæ*
are abundant, mingled with a host of amorphous particles too greatly comminuted to be referred to any
determinate form. Enough is seen, however, to show
the organic origin of the flint-masses; and as to the
question of their introduction into the chalk, *that* no
longer remains a mystery.

Among the organisms found in the cretaceous flint
nodules, none have elicited more discussion than cer-

tain bodies named *Xanthidia*. These present some
diversity in shape; but their general form may be
compared to a ball stuck full of pins, each of which
has, instead of a head, an extremity split into three or
four points, which are hooked downward. Ehrenberg
supposed these to be distinct animals, to which he
gave specific names; but they are now known to be
the *sporangia* (or seed-bearing vessels) of certain
microscopic plants, the *Desmidiaceæ*. How these
came to be mingled, in the flints, with products
exclusively marine, was the wonder, since it was believed that the Desmids were never found except in
fresh waters.

Dr Wallich finds, however, *Xanthidia* among the
alimentary contents of pelagic *Salpæ;* with the endochrome so fresh as to make it manifest that they had
recently been taken into the stomach; and this far
out in the limitless ocean. And even adult *Desmidiaceæ* have been found in the same circumstances; so
that the whole difficulty of the association of these
sporangia with marine siliceous organisms vanishes by
the discovery that this class of plants is also marine.

What beautiful chains of mutually dependent links
are presented to us in these investigations! How true
is the aphorism that in the works of the all-glorious
God nothing is great, nothing is small: or, rather,
the small is great; nay, sometimes, as here, the least
is the greatest. Take away the invisible Diatom and
Foraminifer from the ocean, and what would be the
result? Man would not be cognizant that anything

had disappeared ; since his experience of six thousand years has left him utterly unconscious, till yesterday, that such things existed. Yet how soon would the tale be told! and how sadly! What blanks would presently be seen! what great rents in the beauteous web of nature! What distortions of the admirable unity! What disturbances of the delicate balance of creation! The "foundations" of the physical world would be, like those of the moral, "out of course;" and unless some countervail were quickly applied by the remedial wisdom of Him who is infinite in resources, the whole cosmical system might be hopelessly deranged. The whole race of *Salpæ*, and *Ascidiæ*, and *Conchiferous Mollusca* would starve and disappear; entire genera of fishes would be lost; the sea-fowl would starve; the seals and dolphins would perish; the Arctic bear would seek in vain for food; and the great whales would pine and die of hunger. The solitary ocean would be a waste of death; animal life would cease throughout its expanses; the *Algæ* would grow and grow till they had exhausted the carbonic acid, and then die for want of a fresh supply. Putrid exhalations and morbific miasmata would sweep over the land, and death would soon reign undisputed here. What disturbances of existing laws might ensue from the failure of the present incessant depositions of inorganic matter on the sea-bed, we cannot even conjecture; but doubtless these would not be few or unimportant. On the whole, dimly as we discern the catenation of cause and effect, it seems not at all ex-

travagant to presume that all this mundane creation
is actually dependent, for its sustentation in being, on
the existence, in health and abundance, of an animal
and a plant far too small to be seen by the human
eye to which it is presented.

Thus we see how one great function of Divine be-
nevolence, " He openeth his hand, and satisfieth the
desire of every living thing" (Psalm cxlv. 16), is an-
cillary to another putting forth of might by Him who
is our " God and Kinsman,"—" who upholdeth all
things by the word of His power." (Heb. i. 3.)

———

The ink of the above lines was scarcely dry, when
information was received from the deep sea, setting
all our speculations at defiance, and confounding all
our conclusions. *Animal life is actually flourishing
under the pressure of a mile and a half of superincumbent water.* H.M.S. *Bulldog*, under the command of Sir Leopold M'Clintock, has returned from
surveying the Northern Atlantic, from Cape Farewell to Labrador, and Dr Wallich communicates to
the *Annals and Magazine of Natural History* the
following statement, the interest of which will warrant my citing it *in integro :*

" During the recent survey of the proposed North
Atlantic Telegraph route between Great Britain and
America, conducted on board H.M.S. ' Bulldog,' some
important facts have revealed themselves, from which

it would appear that all preconceived notions as to the bathymetrical limits whereby animal life is circumscribed in the sea are more or less erroneous. The mighty ocean contains its hidden animate as well as inanimate treasures; and it is probable that, under proper management, the former may speedily be brought to light, whatever may be the ultimate fate of the latter. In short, we are almost warranted, from the evidence already at our command, in infering that, although hitherto undetected, a submarine fauna exists along the bed of the sea, and that means and opportunities are alone wanting to render it amenable to the scrutiny of the naturalist.

" In sounding midway between Greenland and the north-west coast of Ireland, at 1260 fathoms—that is, at a mile and a half below the surface, in round numbers—several *Ophiocomæ* were brought up, clinging by their long spinous arms to the last fifty fathoms of line. They were alive, and continued to move their limbs about energetically for upwards of a quarter of an hour after leaving their native element. The species seems allied to *O. granulata* (LINK), the specimens varying from two to five inches across the rays. Lest it be supposed that these *Ophiocomæ* were floating or drifting in the water at any point intermediate between the surface and bottom, it is only necessary to mention, that the determination of depth having been effected by a separate operation and apparatus, the more tedious process of bringing up the sample of bottom is entered on; and, owing to the difficulty of finding out the exact moment at which ground is struck, a considerable quantity of line in excess of the already ascertained depth is usually paid out. This quantity, therefore, rests on the bottom for a short time, until the sounding machine is again hauled up. The *Ophiocomæ* were adherent to this last fifty fathoms only, and were not secured at all by the sounding machine. It is quite clear, therefore, that they were met with on the surface-layer of the deposit. The distance from the nearest point of Greenland to the spot at which this sounding was made is five hundred miles, and to the nearest point of Iceland (namely, an isolated rock called the 'Blinde Skier,' about seventy miles from the mainland) two hundred and fifty miles; so that, admitting the possibility of the starfishes having been drifted by currents, for argument's sake, the character of the fact would be in no way affected. The structure and habits of the Echinoderms generally are too well known, however, to render such a mode of accounting for their presence in the position referred to possible.

" On careful dissection, I found no appreciable anatomical difference between these *Ophiocomæ* and the species frequenting shoal waters. The deposit on which they rested consists of [certain *Foraminifera*, named] *Globigerinæ*, so pure as to constitute ninety-five per cent. of the entire mass. Their occurrence where the *Globigerinæ* are to be met with both in greatest quantity and purity, together with the circumstance that in the stomach of the *Ophiocomæ* the

Globigerinæ were detected in abundance as alimentary matter, corroborates the evidence I have obtained from other facts as to the normal habitat of the latter organisms being on the immediate surface-layer of the deeper oceanic deposits, and not in the substance of the superincumbent waters. At the same time it substantiates the truth of the star-fishes having been captured on their natural feeding ground.

"I also detected, in a sounding made at 1913 fathoms, a number of small tubes, varying in length from one-sixteenth to a quarter of an inch, and about a line in diameter, which, on being viewed under the microscope, turned out to be almost entirely built up of young *Globigerinæ* shells, cemented side by side, just as we find to be the case in the tubular cells of some of the cephalobranchiate Annelids, where sandy or shelly particles are employed in their formation. There can hardly be a doubt, therefore, that some minute creature, probably an Annelid, lives down at this enormous depth, and feeds on the soft parts of the Foraminifera, whilst he houses himself with their calcareous shells. As yet I have been unable to determine the nature of these creatures, but hope to be enabled to succeed on a more lengthened survey of the material in which they occur.

"Lastly, I would mention having met with the minute bodies termed 'Coccoliths' by Professor Huxley. They occur in vast numbers, associated with larger cell-like bodies, on the surface of which Coccoliths are arranged at regular intervals, so as to

lead to the inference that the latter are in reality
given off from the former in some way. The larger
cell-bodies, and the Coccoliths on them, are imbedded
in a gelatinous envelope. The presence of these
organisms in largest quantity in those deposits in
which the *Globigerinæ* occur alive in the greatest
profusion and utmost state of purity, would also seem
indicative of their being a larval condition of the
latter."

As the supposition that the pressure of so great a
body of water would preclude the possibility of animal
functions being carried on at the bottom of the ocean,
is thus found to be a mistake; so it is by no means
improbable, that our received theories of absolute
darkness at that depth may be equally mythical.
Edward Forbes formed an ingenious hypothesis touching the distribution of marine animals in zones of
depth, from facts which seemed to prove that positive
colour diminished in the shells of the *Mollusca*, in the
ratio of their habitual distance from the surface, all
colour ceasing at from fifty to one hundred fathoms.
It was hence assumed that light was entirely lost by
absorption, in passing through such a volume of sea-
water. Subsequent researches, however, by Sars, and
other Norwegian naturalists, proved the existence of
certain Anemones and corals at a depth of two hundred fathoms; and these are by no means white, as
this hypothesis required, but adorned with the most
vivid hues. Light, then, must exist, and have a strong

M

colorific power at that depth.   Dr Wallich has not
alluded to the colours of his *Ophiocoma ;* but as he
compares it to *O. granulata,* we may fairly assume
that there was no great disparity in hue.   Now this
species is of vivid colours :—black, brown, orange,
roseate, are the tints of the disk ; and that of the
rays, dusky white, or bluish.   Can the colour-pro-
ducing rays of the sun, then, penetrate through a
stratum of water a mile and a half thick ?   " No ;"
say the philosophers, " absurd ! "   " Yes ;" says the
Ophiocoma, " ecce signum ! "

Highwater Mark.

HIGHWATER MARK.

THE CORBONS HEAD, TORBAY.

THE horizontal gleam of a December sun, gilding the
wall of our breakfast-room, stirs our valorous heart to
forsake the amenities of fireside and arm-chair, and to
face the sharp morning air. And we are the more
powerfully incited to the deed of daring, because it
is the first fair beam of sunshine that we have been
favoured with for at least a week, during which period
a raw easterly hurricane has been blowing keenly and

fiercely; and each day more keenly and fiercely than on the preceding.

But there are indications of a change. The wind has veered to the north; and, if it is really colder, our feelings give the lie to the thermometer, for it does not feel half so cold. The leaden dreariness of the sky is breaking up into hard mottled clouds, and there is a bright belt of transparent gold that underlies the whole all round the eastern horizon, which augurs well for the day. We go to the garden-gate, look stedfastly in the wind's eye with clenched teeth, button up our coat, and—are off.

December though it be, it is Devonshire; and as soon as we have got well clear of the high road, and have turned into a narrow winding lane that leads straight down (as straight, that is to say, as a Devonshire lane *can* lead) to the sea, we have forgotten both cold north wind and warm fireside.

The banks rise high on either hand, crowned with yet loftier hedges, like sheltering walls. Many a red and brown leaf still hangs on the brambles, and the glossy ivy creeps and twines among them in a close mat of verdure, uninterrupted for rood after rood, ever and anon towering above the hedge in a dense bush; or climbs and fills the naked oaks and elms, spreading wide its umbels of pale blossoms, or of newly-formed green berries. But below the level of the ivy, how rich and varied a mass of verdure yet defies the winter storms! The rose-campion and the herb robert still show their crimson blossoms; and the curiously-cut foliage of this latter and of the shining crane's-bill attract the eye, varied with the fleshy coin-like leaves of the pretty navel-wort in great abundance; while, over all, arch and droop, in the most gorgeous profusion, enormous tufts of that most elegant fern, the hart's tongue, whose long glossy fronds of richest green afford the best imitation of those glories of the tropical forest, the *Musaceæ;*—an imitation on an humble scale, indeed, but yet sufficient to recal with vivid recollection, to one who has seen them, the appearance of those noble leaves, as they break out of the dense mass of forest foliage, and droop on either side of some narrow bridle-path in the mountains of Jamaica.

Nor less inviting is the soft and tender verdure of the mosses. It is their season of rejoicing. In

"——The time of flowers, the summer's pride,—"

these frail beings wither and dry up; but under the fogs and rains of autumn, and the winds and frosts of winter, they spring to new life and vigour. Recovering all their beauty, they spread in soft fleeces of verdure, and shoot up their slender stalks, crowned, as here, each with its tiny urn, and wearing its fairy nightcap. Look at this flat stone, draped all over with brilliant *Bryum!* Surely the fairies must be here holding their "board of green cloth;" only our dull, prosaic eyes are not worthy of beholding them. And see, everywhere around,—on the stones, on the summit of this ruined wall, on this decaying tree-stump,—are the little round velvet cushions of *Tortula*, the seats, doubtless, of the august assembly!

The gaping capsules of the fetid iris are displaying here and there their orange seeds crowded within like glowing coals of fire; the crimson haws, and the scarlet hips of the dog-rose hang thick on the thorns and briers; and here, under the shadow of the ferns at the bottom of the bank, shoots up from the damp moss a tiny vermilion agaric. We must take a closer glance at this. How delicately tender! The most cautious touch of our fingers crushes its succulent stem: but what a beautiful little conical cap! bright scarlet without, like coral; and within, oh! how perfectly beautiful the order and arrangement of these radiating plates! We pluck away the stalk, and then, if we did not know the origin and texture of the object in our hand—if we saw it under a glass shade—we should not hesitate to affirm that it was actually a madrepore, one of those cups which we may find affixed to the caves of yonder shore, the concavity of which is lined with plates of stone, the very counterparts of these! Here are the several cycles of *cloisons*, each intercalated according to its subordination, exactly as we see them in the zoophyte. So curiously has the wisdom of God repeated, as it were, in remote regions of creation, the same idea of grace and beauty! Nor is this by any means a singular example.

Now, emerging from our winding mossy lane, the sea in its boundlessness suddenly breaks upon us.

We have but to cross a high-road and we are on the beach: so sudden is the transition from the intensely rural to the maritime! Now once more we feel the furious northern gale; but we are warm with our walk, and defy it. A moment's pause to take in this characteristically wintry prospect of the sea. Beautiful is the ocean at all times; most sweetly beautiful when it sleeps, stretched out in silvery brightness, "like a molten looking-glass" under the azure sky of summer; but most grand, most full of majesty and power, when, as now, it chafes and foams beneath the lashing gales of winter. Ha! winter is a sterner schoolmaster than the Persian. The rollicking Euxine laughed at his chastisement; but stern Boreas knows how to lay on the lash, till the writhing element shrieks, and roars, and groans under the infliction.

The gleam of sunshine is gone, and the sky has settled down again in frowning gloom. A black and threatening brow it wears; and the well-whipped ocean—tortured but unsubdued—looks up with an equally threatening blackness, save where the thousand crests of foam rise and fall, tossing and careering on their rapid shoreward course.

How fast they chase each other on, as if eager to escape the furious strokes of the driving breeze behind! And when they reach the friendly strand, how each in quick succession gracefully rears its green glassy wall, curves-over its crest, and pours its long cataract of foam high on the yellow sand! A beautiful sight; but in a moment an abject ruin is all that remains, and in another this is covered and obliterated by its successor.

It is a wide indentation of the Channel coast on which we are gazing. On the left hand, lofty wooded hills, covered with the white suburban villas of a flourishing watering-place terminate in sloping points of rock, off which two rugged islets lie like chained lions guarding the port. To the right, the view is similarly bounded by a long nearly level wall of high down, ending abruptly in a bluff and perpendicular headland. Between these expands a long range of angry horizon, and at our feet stretches for a couple of miles a beach of yellow sand.

Well, now, have we seen all of interest? and shall we go back? By no means; we have come out exploring, and we have only just reached our hunting-ground. But is there anything to be found on this naked beach, on which the sea is breaking so furiously that we cannot approach the line of low-water? Yes, much; if we only know where and how to look for it.

Do you see a long black line, or belt, a yard or so in width, which your eye may trace along the whole length of the yellow beach, lying parallel with the sea-edge, but far up, almost close to the landward verge of the sand? This is the sphere of our operations to-day.

This is the line of high water; the mark to which the waves have reached at the highest flood-tide, where they have deposited the spoils which they had collected from various sources, and which they had

borne on their bosoms hither. Each careering billow
carries on its summit all the floating *rejectamenta* of
the sea; and, as it rushes in its fulness on the beach,
pushes these trophies of its prowess in its van. But
there it leaves them; for it retires, not in a full flow,
as it came, but grovelling downward among the
gravel, and partly sinking into the sand; the *spolia*
arrested on the sand, to be thrown higher if a higher
wave should come and lift them, but never to be
returned whence they came. Here they remain, the
flotsam et jetsam, which our old maritime laws assign
to the Lord High Admiral as the perquisite of his
office—make what he can out of them.

Well, as he is not here to look after his own, we
will take the liberty of the first search. Fortunately,
he won't grudge what we shall take; for, as Crabbe
says of his insect-hunting weaver—

"Ours is untax'd and undisputed game."

A moment's glance at our feet suffices to show why
this belt of various materials is black. For nine-
tenths of the mass consist of the coarser seaweeds of
the *Melanosperm* order; chiefly the wracks and
tangles, the *Fuci* and *Laminariæ*, which, though
olive or brown while living or fresh, speedily become
black when their surface dries. The profusion with
which such plants line the beach after these winter
gales, shows the great force of the sea; for the waves,
though only agitations of the surface, the deep water
being waveless, extend at low tide to the great forests

of olive sea-weeds that fringe the rocks at and below
the lowest tide-levels, and tear them up from their
moorings to cast them thus high and dry on the beach.

The force thus exerted you may better appreciate if
you have ever tried to pull off living specimens of the
common tangle. The strongest man may pull and
tug in vain ; though the stout and rough stem affords
a capital purchase for the exertion of his muscular
powers. A full-grown tangle, such an one, for
example, as this at our feet, with a stem an inch in
diameter, would probably mock his most strenuous
efforts. The rock itself will frequently give way before
the attachments of the weed.

This very specimen shall be honoured with our first
observations ; nor shall we find it unworthy of our
attention. It is the common tangle, or fingered
Laminaria, which grows in great abundance all round
our rocky shores, forming a broad belt of dark waving
submarine forest, of which the summits are just
exposed at the lowest recess of the tide. It consists
of a root, a stem, and a frond. The root is a remarkable structure, and instructive as an example of the
perfect manner in which the creative wisdom of God
achieves the same end by different means. Stability
is secured to the forest-tree by the repeated subdivision and wide ramification of its roots, which penetrate
into the soil ; the sea-weeds, on the other hand, do not
penetrate the soil, and have no true roots. This tangle
grows on the solid impenetrable surface of the rock ;
and the problem is, how to impart to it stability of

attachment. The base of the sea-weeds is a simple
adhering disk; but since, in this case, the great
expanse of the frond gives a great advantage to the
force of the waves, the adhering disks are made very
numerous, and are spread over a considerable area.
We see the bottom of the stem dilated into a conical
mass, some four inches in diameter, of smooth, rounded
branches, which are stout where they diverge from the
stem, but ramify and re-ramify at very short intervals,
until they produce a crowd of slender but firm fibres,
each of which terminates in a flattened expanded base
or disk ; thus taking hold of the rock in a multitude
of separate attachments, and forming, when combined,
a very strong adhesion.

The stem which springs from the summit of this
root-cone is about an inch thick, but gradually tapers
to half that diameter, constituting a straight round
rod, becoming somewhat flattened towards the top.
Its surface is rough when old, its substance firmly
flexible, somewhat gelatinous, within. You would
scarcely expect to find a substitute for buck-horn in
this slimy sea-weed ; but in some parts of our coasts
serviceable knife-handles are made of it. A pretty
thick stem is selected, and cut into pieces of suitable
length. Into these, while fresh, are inserted blades of
knives, such as gardeners use for pruning and grafting.
As the stem dries, it contracts and hardens, closely
and firmly embracing the hilt of the blade. In the
course of some months the handles become quite firm,
and very hard and shrivelled ; so that when tipped

with metal they are hardly to be distinguished from hart's-horn.* If you are disposed to try the experiment, for I cannot say, *probatum est*, I would recommend that the stem should be well soaked in fresh water, to avoid the unpleasant effects of the salt alternately drying and deliquescing.

At the summit of this stem we see what we may fancy to have originally been a great piece of well-curried calf-skin, some three or four feet broad in every direction. But this has been irregularly split into straps of varying width, almost as far down as the union of the plate with the stem; and the extremities of these divisions are, in such a specimen as this, rudely torn and jagged. The surface, however, is beautifully smooth and glossy, of a rich dark-brown hue; the texture is firm and tough, and yet so flexible, that we cannot help thinking it a pity that it has not yet been turned to account by any enterprising cordwainer of the "*pannus corium*" vein. Buckhorn from the stem, and buckskin from the leaf, would be a pretty double manufacture from the "*alga vilis.*"

But in picking up a great tangle like this, we find much more to instruct and delight us than the actual plant. The wonderful principle of parasitism which pervades nature is full of interest. Perhaps the poet assayed a somewhat loftier flight than our present observations warrant, when he, too rigorously, asserted, that—

* Neill.

" Great fleas have little fleas
Upon their backs to bite 'em;
And little fleas have lesser fleas;
And so *ad infinitum.*"

But still it is true that the sphere of life is immensely
augmented by this remarkable device of making one
organism a microcosm on which other organisms
grow and revel; making this stem, for example, a
region on which forests of other plants may wave,
and this strap a plain on which an enterprising poly-
zoan may build a populous city.

We break off, with some excoriation of our fingers,
the outermost of these tough rootlets; and discern
that their conical contour encloses a smooth-walled
chamber, sufficiently capacious to afford " ample room
and verge enough" for the residence of a luxurious
epicure, who has an oriental repugnance to locomotion. How snugly ensconced is this overgrown limpet!
You wonder how ever he got in, and how ever he was
to get out. The fact is, he got in a long while ago,
when he possessed the slenderness of youth, before
many of these rootlets were formed; and as to getting
out, that contingency never entered into his brain (I
beg his pardon, he has no brain; well then, into his
cephalic ganglion). What can a mollusk want more,
when he can feed on the wall of his bed-chamber,
and finds the savoury nutriment grow faster than he
can lick it off? It would seem, indeed, as if the exertion of roaming over these narrow walls were too
great for him; that, so far from complaining of the

res angustæ domi, the greater part of his dwelling remains a *terra incognita* to him; for, when we drag
him from his foot-hold, we find that he has really
lived in a cavity commensurate (and no more) with
the outline of his body; as if the substance of the living wall had gradually grown around him since he
first set foot there, just as the soil of London has
risen around the site of the ancient St Paul's.

And so it has, doubtless. Even though a more
intimate acquaintance with our eremite's manner of
life should induce us to reject the supposition of his
actual immobility, there can be no question that
this cavity is fashioned on the animal's body, partly
by a slow erosion of the surface, and partly by the
growth of the plant. And yet he may occasionally,
nay, frequently and periodically, wander. Some
curious passages in the history of his bigger brother,
the common rock-limpet, may help us to a little
light on the matter. We often find this familiar
species imbedded in a shallow, but very perceptible
and well-defined, depression *excavated out of the solid
rock*,—shale, or limestone, or whatever else it may
chance to be; but this cavity in every case so accurately fitting the dimensions of the animal, as to
leave no room for doubt that the former has been
really modelled by the latter. Now the limpets are
all exclusively vegetarians in their diet; and the
problem was to understand how the creature could
procure his daily dinner of greens, while yet he was
so manifestly a fixture. But a peering naturalist,

out one night upon his sea-side prowlings, with a
bull's-eye lantern at his girdle, much to the mystifi-
cation, no doubt, of the Coastguard watching on the
cliffs above, observed the strange phenomenon of old
aldermanic limpets crawling hither and thither with
tilted shells, about the tender mossy green that grew
in patches on the rocks. He was curious enough to
mark their movements ; and found that as morning
approached the limpets, in a comfortable state of re-
pletion, glided away from the mossy patch, and
betook themselves, with unerring precision each to
his own hollow in the stone, into which he settled
himself as snugly as if he had never moved at all.
Perhaps our friend of the slimy chamber may have
similar instincts.

There is not much of beauty to recommend this
species to us ; the shell shows its successive additions
with uncouth distinctness ; and the later of these are
rough and coarse, and seem scarcely congruous with
the smooth and prettily-painted apex. But, as with
some higher organisms, beauty is the endowment of
youth ; age plays sad havoc with personal attractions,
in limpets no less than in ladies.

Let us look for one of the rising generation of
limpets. We must search, not in the root-chamber,
for they do not take to house-keeping till they have
attained a certain age. Here, abroad on the free
pasture of the leathery leaf, we may hope to find
them. Yes, here are several, scattered over this
smooth olive strap. But how different from the

N

closeted octogenarian! The shell is of unimpeachable symmetry, polish, and delicacy; it is of a translucent horn-colour, and its summit is marked with
three fine lines of the most brilliantly-gemmeous
azure.

But we have not quite exhausted this mass of tortuous roots; for here, peeping from their narrow
interstices, we discern two or three tiny knobs of
brown flesh, which, shrinking from the touch, manifest their animal nature and their vitality. We can
make little of them in their present condition; but,
breaking or cutting off the rootlets which hold them,
let us drop these into a phial of sea-water, and we
shall see—what we shall see.

The minute knobs of flesh are beginning to swell and
protrude from their crevices, as they feel the geneal
stimulus of the water, and now they form little cylindrical columns with rounded summits. Now those
summits are opening; from a central point issue tiny
filaments, at first as a little crowded pencil or *fascis;*
but, as the opening expands, these also recede until
at length they stand, a crown of sensitive tentacles,
around the margin of the short pillar.

And the apex of the pillar, the tiny area surrounded and walled-in by these environing guards,
how beautifully is this decorated! Like some gay
pattern which a child has caught in twirling a kaleido-
scope, and calls on his mother to see and admire,
we behold here a shallow saucer painted in a many-
rayed star of yellow, and orange, and pink, of purple,

and brown, and white; while the fringe of moving
tentacles is varied with rings of white and dusky upon
a pellucid gray ground. Surely, for variety and rich-
ness of colouring, minute as it is, this little creature
deserves the title which has been assigned it, of the
Ornate Anemone.

The vivifying element has revealed another of the
same tribe, which we had failed to detect before. Of
similar form and dimensions, it is arrayed—*simplex
munditiis*—tentacles, disk and column in virgin white,
the purity of the *undriven* snow, of the snow that has
fallen silently and impalpably throughout a calm and
breathless night.

These are *Anemones;*—things which we are accustomed to put at the bottom of our zoological scale;
things which we call zoophytes, animal-plants, ani-
mal-flowers; and of which our fathers thought, as
many of their children still think, that the vegetable
element predominated over the animal. Sluggish,
flabby, helpless creatures they seem; ready to become
the unresisting prey of the first fish that espies them
and essays to taste their daintiness. Yet let him be-
ware! He will rue the hour when his wretched lips
touched the fatal morsel, should he be rash enough
or young enough to make the venture; but indeed
instinct is generally a teacher potent enough to induce him to practise the *laissez faire* policy in this
case.

Here is this Snowy Anemone expanded to the full
in all its unsuspecting bridery. We give it a poke

with the end of this straw. Instantly its charms are
turned in, and packed away from the rude insult, and
the pillar begins to shrink into a round button, and
to retire within the friendly shelter of the encompass-
ing roots. But lo! as it goes a fine white thread
shoots out from the surface of the body to the length
of a couple of inches; another, and yet another! half
a dozen threads dart from so many points, and stream
away through the clear water, or intertangle with
each other, or curl up in spiral coils and irregular
contortions.

It is thus that the anemone seeks to avenge the
affront that it has received. These threads are so
many valiant and skilful warriors posted at the battle-
ments and loopholes of her castle; or rather her light
horse, which, issuing from a hundred sally-ports,
scour the surrounding region in search of the insolent
foe. And woe betide the foe, be he slug, shrimp,
worm, or fish, who incautiously comes within the
range of these archers! they will pierce him through
and through! and their shafts are poisoned with as
fatal a venom as ever issued from the unerrring Indian's gravatàna on the banks of the Amazon.

These apparently simple threads, that you might
well mistake for fragments of sewing-cotton dropped
by a lady's scissors, are master-pieces of ingenious contrivance and wondrously-elaborate mechanism. Death,
nay, myriads of deaths are sealed up in every inch of
this thread; certain, inevitable death. It is the veri-
table thread of the *Parcœ*, to the tenants of the waters.

Nothing is more wonderful than the structure of this thread, when we examine a minute portion of its length by the aid of the microscope. It is made up of millions of transparent oval or oblong sacs, bags of clear, tough membrane closed at each end, and of such dimensions that five hundred of them, if placed in contact end to end, and three thousand, if laid side by side, would lie within the length of an inch. Within each capsule is seen a wire coiled up loosely, and thus occupying the interior.

Such is the armature as it lies ready for action. But, on the stimulus of the animal will, the oval capsules project from the periphery of the containing thread by thousands, and instantly each shoots its shaft. In other words, the coiled-up wire is evolved with lightning-like rapidity from the smaller end; and that not by direct projection, but by eversion; for, wonderful to tell, the wire is tubular throughout, and in order to be shot it must be completely turned inside-out, just as you invert a stocking.

Well, you see this subtle wire running out, evolving its inner surface as it goes, so quickly that it is only under rare circumstances that the eye can follow the process. But now you discern that even this wire is not a simple tube. For throughout its entire length—a length which often reaches to thirty times that of its capsule—it is ridged with one, two, or three elevated *carinæ*, or thickened bands, which run round it in regular spiral turns, at exactly prescribed distances, and at an exactly prescribed angle of inclination. These thickened ridges carry a number of fine stiff bristles, which, in the quiescent condition of the coiled wire, lay close in its interior; but as soon as the progress of the evolution frees them, fly out, and presently assume a retrograde direction as so many reverted barbs.

The force with which this remarkable wire is shot is sufficient, combined with its almost inconceivable tenuity, to enable it to pierce the skin and tissues of the animals with which it comes in contact. A bit of fish-skin, examined microscopically after an instant's touch by one of these threads, is seen to bristle with barbed wires, like a target after a day's archery.

Moreover, in some way or another, a most subtle poison accompanies the evolution of the wire, and is injected into every wound made by its intromission. No animal of small size, however vigorous its life, however exalted its rank in the scale of being, can withstand its power. It is most fatally destructive. A few seconds in general, a few minutes at most, suffice for the withering up of its activities; torpor quickly sets in, and quickly death. "Nemo me impune lacessit" is the device of the ancient house of Ἀκαλήφη.

Did I not truly say that we have here a most elaborate piece of mechanism? And surely its wondrousness is greatly enhanced by its minuteness! If we admire the skill of the penman who writes the Decalogue in the area of a threepenny bit, and the Iliad in a nutshell, though with no ulterior end, what shall

we say to the skill which forms engines of battle,
such as these, and packs them by millions in an inch
of thread, not for the useless display of power, but for
the defence and sustentation of creatures which Om-
niscience has devised and Omnipotence has created?

The naturalist has learned to see beauty where
others see ugliness; nay, he can see what puts him in
raptures, where the uninitiated eye discerns nothing.
It is curious to note the difference between the in-
structed and the uninstructed sense; between the
perception sharpened by habitual use, and by strong
desire, and the same perception in its ordinary, and
what we may call its *passive* exercise. The latter
takes in a wider range, but the impressions it receives
are proportionally vague, and their duration brief;
while the details altogether escape notice. The former discerns little, except what bears upon its immediate object; but within that sphere nothing escapes
it. Instead of the wide but indefinite diffusion of
the perceptive faculty, there is here the concentration
of it upon some object or series of objects, which are
discerned with vivid intensity, and a corresponding
isolation from all other, irrevelant, objects.

A remarkable example of this combination of concentration with isolation, is mentioned in Lander's
" North of Europe." " A man set off one morning
to shoot the Tjader, or Cock of the Woods, which is
effected in this wise: the bird is so extremely shy,
that he may rarely be met with, except in the pairing

season, when every morning he renews his song.
He usually commences just before sunrise, beginning
in a loud strain, which gradually sinks into a low
key, until he is quite entranced with his own melody :
he then droops his wings to the earth, and runs to
the distance of several feet, calling ' Cluck ! cluck !
cluck !' during which time he is said *to be incapable
of seeing,* (so wrapt up is he in his own contempla-
tions,) and may be caught even with the hand, by
those who are near enough. As the fit lasts only for
a few moments, the sportsman must, if unready, wait
for the next occasion ; for, should he advance a step,
except when the bird is thus insensible, he will certainly be overheard, and the victim escape.   The
man I began to speak of, being early one morning in
pursuit of the bird, heard his song at a short distance ; and, as soon as the clucking commenced, of
course advanced as rapidly as he could, and then
remained motionless till those particular notes again
sounded.   It was quite dusk, the sun not having yet
risen ; but the song seemed to come from the centre
of an open space in the forest from which the sportsman was just emerging. He could not see many
yards before him, and only followed the direction of
the sound. It so happened that, from another point,
but at no great distance, a bear was advancing on the
Tjader, just in the manner of, and with the same
steps as, the man. The hunter, while standing motionless, thought he perceived a dark object on one
side of him, but *it did not much engage his attention;*

and at the usual note he moved on towards the game, but was surprised to see that the black object had also advanced in an equal degree, and now stood in a line with him. Still he was so eager after the bird that *he could think of nothing else*, and approached close to his prey before he perceived that a large bear stood within a few feet of him ; and, in fact, just as they were both about to spring on the bird, they caught sight of each other, and each thought proper to slink back. In this case, the bird, the man, and the bear were all rendered insensible to impressions which at other times would have been instantly perceived, by the very intensity with which the senses of each were concentrated on one object. With the Tjader it was his mate, with the man it was the bird, with the bear, ditto."

The power which the senses acquire by this sort of intense and isolated exercise often appears wonderful to those who have never perceived otherwise than passively. Civilised men marvel at the ease and precision with which savages make their way from point to point, through dense forests, across wide plains and prairies, over rugged mountains, and along deep valleys, and by night as well as by day. And some have even surmised that the savage man has a sixth sense, unknown to us. But it is only the result of exercise and concentration. The white men who follow the occupation of " trappers," in the defiles of the Rocky Mountains, acquire the same perceptive powers as the Indian, the same habit of drawing conclusions from

the minutest, and to others altogether imperceptible,
phenomena; a broken twig, a crushed leaf, a bent
blade of grass, a slight depression in a bed of moss,
the action of a distant bird; all are inevitably discerned, and each has its eloquent record to their
minds. These feral white men, too, are able to travel
in direct lines without sun or compass.

Strange to say, the perceptive faculty, in its highest
condition of exercise, seems to revert again to the passive state.   These men of the wild, whether white
or red, at length come to exercise their acute powers
without effort and without consciousness.   The desired result is infallibly attained; but when asked by
what means, they cannot tell you.   They have not
been conscious of the individual processes by which
they formed their conclusions.   Take, for instance,
the travelling through a dense American pine forest.
Certain indications have been observed, by the knowledge of which the points of the compass may be determined, such as the condition of the bark of trees,
the mosses and lichens, which always grow thickest on
the north side, the direction in which the summits of
certain conical trees, as the hackmatack, and some
others of the pine family, bend over, which is invariably
to the north-east.   But the Indians protest that they
do not have recourse to these or the like signs.
Hardy,* who has investigated the subject with some
care, thinks that they really do not.   When he has
mentioned them to a Red Man, he has invariably

* Sporting Adventures in the New World.

laughed heartily, saying, "Indian no want look at bark or tree-top,—'cept when he hunt porcupine." But if the Indian had had as acute a power of analysing his perceptions as of forming them, no doubt he would have found that he had been every instant receiving and treasuring impressions from such phenomena, though the process had become so habitual that he was now unconscious of it.

But we have travelled far from High-water Mark; and you are wondering what may be the text of this long sermon on instincts and perceptions, red men and pine forests. It is indeed a small one, a tiny gray scale, not so large as the diameter of a split pea, spread upon the stem of this sea-washed Oarweed, which almost every one would pass by as nothing; but which, to you and me, fair reader, because we are naturalists, is a volume of biography.

Disregardful, then, of that young gentleman with the eye-glass and cigar, who gives us first a supercilious, and then a pitying glance, and looks anxiously hither and thither, no doubt wondering where our "keeper" can be—let us sit down on this old spar, and read our little history.

By the aid of a pocket-lens, then, we see a tiny plate of glistening white shell, of a roundish outline, adhering to the rough foot-stalk of the weed. We can pass the tip of a penknife under the edge all round, and we see that the point of adhesion is only the centre of the base. The upper surface presents a complicated structure; for to a short distance within

the margin, all round the circumference, it is plain, like
the border of a dinner-plate, but marked with radiating bands, which are alternately opaque and pellucid
white.   The area of our fairy plate is occupied with a
multitude of short tubes formed of the same white
shelly substance, which crowd closely on one another,
and project obliquely outwards and upwards from the
centre, in regular succession, forming radiating series.
Thus we can look down into the tubes, the circular
apertures of which are open; but those nearest the
centre of the plate are much shorter, and can scarcely
be considered as tubes, but are rather shallow
cells.

The tubes are empty now; they are like the houses
of London after the plague; whole streets tenantless,
because the inhabitants have died out.   But hold!
here is another tiny fragment on the root of the same
weed.   Let us examine this.

It is much like the little china plate, but the mass
is more spreading, and projects in irregularly sinuous
lobes, like the outline of a coast in a map, and is of a
pale lilac hue.   Its surface is studded with glassy
tubes, which are larger than the former, and are prettily arranged in short rows, with considerable space
separating row from row.   Three, four, or five tubes
start from the surface and stand obliquely upwards,
soldered together by their sides, forming a single row;
then, a little way off, three or four form another similar row, and so the whole surface is covered.   But the
edge of the plate is composed of tubes much crowded,

and set horizontally, so that their apertures project on the margin.

And here is another form still. A narrow band of shelly matter, not much thicker than a pin, creeps along the root of the weed, and then divides into two spreading branches. The tubes are arranged here in transverse rows along the whole surface, in such a manner that a central line divides them into two sets, those on either hand arching outwards towards that side, so that a furrow separates them.

All these pretty objects—and they are very attractive to the eye when magnified—are closely allied to each other; they are different species of the genus *Tubulipora*. At least, they are the "mortal remains" of such animals, for they are at present mere skeletons: the animals, as I have already observed, have died out. They are exceedingly delicate, and a very short exposure to the air is enough to close their frail lease of life; so that it is rather rare even for the experienced naturalist, who is conversant with such creatures, to get a glimpse of them in healthy vitality.

Each of these tiny tubes is inhabited, during life, by a creature of transparent tissues, whose structure manifests a relationship, not very remote, to the tenants of bivalve shells, such as the oyster or mussel. The tube represents the two valves of the shell, united along the edge, and open at one end; but this assertion must not be considered as anything more than an aid to your conceptions of the affinity of the two forms; for considerable modifications would require to be made before one of these atoms would be transformed into a decent oyster, even in miniature.

One of the most obvious of such modifications would occur in the breathing organs; and as these constitute in our little *Tubulipora* its chief feature of beauty, as well as the most important agent in its maintenance, and as in both animals their structure is highly curious, we may find interest in examining them in detail.

If we take a Mussel, and, inserting a thin penknife between the dark-blue shell-valves, sever the muscular bands which pass from one to the other and form their power of cohesion, we shall display the whole yellow anatomy within. First of all, each valve is lined by a membrane of thin flesh, the edges of which are free, and are cut into festooned fringes. These two leaves constitute the mantle, and are the agents by whose instrumentality the shell itself is made, and from time to time augmented. This process is a very curious one, but, as it is not relevant to our present inquiry, I will not now touch it farther.

Lifting these membranes, then, there appear laid along beneath them two other leaves on each side, thinner and more delicate. These four leaves, which are the breathing organs that I mean to describe, with the two leaves of the mantle, bear the same relation to each other as the leaves of a book, a book of six leaves, of which the two shell-valves are the boards or covers. These gill-leaves at first sight seem mere thin, lax, motionless laminæ of membrane, so fragile that the slightest touch tears them into strips, the rent invariably proceeding directly across their short diameter.

But let us with sharp scissors cut out a portion of one of these leaves, and, laying it on the lower glass of what microscopists call a live-box, cover it with a drop of sea-water, which we then flatten by putting on the glass cover. Now to the stage of the microscope with it, and put on in succession magnifying powers varying from 50 to 250 or 300 diameters. The whole field of view is a scene of the busiest activity : movements of the most exquisite regularity and precision, and of unceasing activity, are going on in every part. A most attractive spectacle is before us, which evokes our delight not less than our wonder.

We see a great number of threads of great delicacy and flexibility, with a tendency to a parallel arrangement; but forming bundles of varying breadth, lying across each other in various directions. Each fragment is composed of more or fewer of the elementary threads, laid side by side in close contact, or *almost* contact. Along these threads we see a very peculiar series of movements. Under the higher powers of the microscope, these are very interesting. Fixing our attention on one thread, and watching it, we discern a multitude of black points running along the thread with a moderately rapid, equable course, all chasing each other with ceaseless perseverance. This is on one side of the thread; but on the other side a phenomenon exactly similar is seen, only that the

running of the dark points is in the opposite direction. For hours and hours, nay even for days, this
incessant chase goes on, never slackened, never intermitted, until the progress of decomposition has destroyed the cohesion of the parts; and even then the
movement may be discerned lingering in the dissolving fragments.

You will not, however, have acquired an adequate
idea of the nature of this curious phenomenon, till
you have made an examination, your judgment
fortified by the facts you have just observed, of
the living and uninjured tissue. Returning to the
Mussel, look carefully over one of the untouched leaves
of the gills with a good pocket-lens, and then you will
see that there is but one thread. The entire gill-leaf *is
formed out of a single thread,* folded to and fro upon
itself, with some hundreds of turns through the whole
length, and that the running points maintain in every
part the same law of motion, their course being up
one side of the thread and down the other throughout.

But these mysterious black points, what are they?
and what the nature of this strange race, which those
tiny chasseurs maintain with so indomitable an energy,
every one striving, and with success too, to be in *at
the death?* It is very hard to persuade one's-self that
they are nothing at all; and when, after a quarter of
an hour's gaze at the untiring objects, your sage microscopic Mentor tells you that they are no objects at all,
you look up to see if he is laughing at you. But no!
the sage is as sober as a judge; and you demand an

explanation. That the microscope should shew you things which you could not see without it, you expected before; but that it should shew things which, after all, are nothing, that you should see with your eyes objects which do not exist, is marvellous indeed. But it is true.

It will take a good many words, I am afraid, to explain this; and some precision of thought on your part to understand it, when it is explained. It may help you, if you recall the familiar sight of a meadow just ready for the mower under the summer breeze. You have often admired the bands of silvery light, alternating with gray shadows, that are ever flitting across the field. What is it that passes over the surface? Nothing. You know that the charming appearance is caused by the alternate bending and rising of the feathery heads of the grass-blossom in succession, as the breath of the breeze sweeps over them. If we imagine all the stalks removed on both sides of a single row of stalks, or, which is the same thing, practically isolate our attention to this single file, the matter will be simplified, and we shall have a very close parallel to the chase in the microscopic live-box.

The gill-thread, all along its two opposite sides, is fringed with a row of minute hairs, so minutely attenuated that the high powers of the microscope, we have been supposing to be used, are insufficient to reveal them in their individuality. These hairs, which are known by the term *cilia*, are endowed with the power of alternately bending and straighten-

o

ing themselves, and of so doing this that the movements of each individual hair shall be in the most
exact harmony with those of its fellows; not that all
bend at the same instant and in the same degree;
but in a rhythmical succession, just as the breeze,
supposing it of exactly uniform force, bows down the
grass stems in rapid succession. The result is a
series of waves: one *cilium* (or stalk of grass) is at
a given moment perfectly erect; the next, at the
same moment, has just begun to bow; the third is
bent still lower; the fourth has attained its extreme
point of flexure; the fifth has resumed the position of
the third, the sixth that of the second, and the seventh
is again erect. Thus these seven *cilia* would limit
the extent of one wave; and the whole line would
exhibit a succession of such waves.

The wave, however, is for no two successive instants
in the same spot; the first *cilium* immediately takes
the position of the second, the second of the third, the
third of the fourth, and so on; so that while the *cilia*
themselves remain fixed, the waves produced by their
alternate flexion and extension perpetually run along.

If, now, you will take the trouble to draw on paper
a number of hairs, equidistant at their bases, but with
their extremities in the different states of flexion that
I have described, you will see the explanation of the
dark running points. Those *cilia* that are passing
from the erect condition to the point of greatest
flexure, have their tips more separated than if they
were all erect; whereas those which are passing from

extreme flexure to uprightness have their tips more crowded. Now, though under the microscope we cannot detect the individual *cilia*, we can detect the effect of this alternate separation and union of many; the former produces a more transparent, the latter a more opaque spot, as its optical expression; and these opaque spots, the crowded part of each wave, are the dark points which seem to perform their incessant and amusing gymnastics.

Every *cilium*, thus, is perpetually occupied in striking the water; and, like a trireme of a thousand oars, the skilful rowers, as we have seen, keep the most perfect time. If the galley were free, these vigorous strokes, making up in cumulative energy what they lack in individual force, would row it bodily through the water. In many aquatic animals, the *cilia* are effective implements of locomotion; but here they subserve no such purpose, but another, still more indispensable to life. They produce, by their lashing action, powerful currents in the water, which is thus driven uniformly to and fro across the gill-leaves, yielding up its precious burden of vitalising oxygen to the blood, which permeates the thread.

And now let us return to our little *Tubuliporæ*. If we had the good fortune to see one of these in its condition of life and activity, we should discern protruding from the mouth of the shelly tube a microscopic coronet of diverging filaments. Some dozen or so of exquisite threads, of what you might suppose to be spun glass, from the transparence and brilliance of their

material, are set around a sort of mouth, and diverge
in the form of a bell, or like a campanulate flower.
The same wonder-working instrument that we had
used before would reveal that these are, in their organisation, the very counterpart of the many-folded thread
which makes up the gill-leaf of the Mussel. Each of
these pellucid filaments is fringed with a double row
of *cilia*, set on the two lateral faces ; and the optical
appearance produced by their action is, as before, that
of running dark points, which hurry down one side of
each filament, and up the opposite. The structure is
exactly the same as in the Mussel, save that in the
Tubulipora the gill-thread is resolved into a few short
threads, which are set in a circular whorl instead of a
flat lamina.

It would be a problem worthy the attention of a
skilful mathematician and dynamician, one conversant
with the resolution and combination of forces,—What
ought to be the result of such a complex array of
currents ? Twelve rods stand in a funnel-like form,
each of which is beset with energetic oars which drive
the water rapidly down one side, and up the other:—
what general motion will be communicated to the
water in the vicinity as a whole ? I know not what
our Newtons or Laplaces would make of the calcula-
tion in their closets, but I know the result that is
attained in fact. The general movement is that of a
whirlpool : a vortex is produced in the water, the
boundary of whose influence is a circle exceeding in
diameter by many times that of the bell of filaments.

The fluid is whirled round and round with a velocity which ever increases as the diminishing spiral curve approaches the centre of the vortex.

Thus, according to a law in physiology, that the lower the rank of an organism in the scale of being, the less of *differentiation* we find, the less of specialty in the assignment of function to organ,—the two offices of procuring food and of breathing, which in the Mussel are performed by two distinct sets of organs, are in the inferior *Tubulipora* appropriated to one. The whirlpool, which ever brings fresh particles of water to the surfaces of the filaments, that the oxygen which they carry may be in succession absorbed by the blood which penetrates into these organs, brings something else. Whatever atoms of organic substance, the dissolving constituents of some decomposing animal or vegetable, whatever roving animalcules, whatever spores of algæ, chance to come within the margin of this Mahlström, they are sucked into its circle, and are then inevitably hurled round in its embrace; at every revolution approaching nearer to that open gulf at the bottom, that living grave, into which, one by one, they descend with a gurgle and a gulp.

And this is how the *Tubuliporæ*, and hundreds more of creatures of like kidney, procure their dinner.

HIGHWATER MARK—(*continued.*)

PETIT TOR.

"First comes David,
  Then comes Chad;
  Then comes the wind as though 'twere mad."

So says the country proverb concerning the saints and
the weather of the March calendar. And surely the
present year has not belied the concentrated wisdom
of the adage, for "wind, and enough of it," we have
had. How it has raged, and raved, and howled!

blowing, to use a seaman's phrase, "as if 'twould tear the very canvas out of the boltrope!" No craft could shew a rag to the gales; and, what is strange, instead of the raw easterly wind usual at this season, it has been all up-channel; balmy and humid, despite its violent fits of squalling. Behold one result in yonder fleet of outward bound, lying under the lee of the bluff Berry Head, a goodly argosy of some two hundred craft of all sizes, from the stately East Indiaman to the queer little French lugger. Well, there they ride, the tall and taper top-gallant mast-heads marking incessant triangles on the sky, and the small fry bobbing about like ducks upon the surging swells that come in from the wild Atlantic. There they ride; or at least they *did* ride, for on looking seaward this morning, they have almost all disappeared, and the shrieking wind is moaning itself to sleep, like a wilful child.

Now to the shore again for the *opima spolia*, the *rejectamenta* of the gale. Down from our height to the sea-beach of the little cove of Petit Tor, through this wild and broken, but verdant combe, between that perpendicular promontory of limestone on the one hand, and yon beetling cliffs of red sandstone on the other. How grand those masses of old conglomerate, from whose lofty brows come the hoarse calls of the choughs, and jackdaws, and where now and again the blue rockdove flits out on swift pinion uttering its loud coo!

'Tis a bright and balmy morning: and as we de-

scend the steep and winding footpath down the dell,
the sheltering hill makes it perfectly calm, and the
full blaze of the morning sun gives to the air the
warmth of summer. Vegetation is rapidly unfolding in so genial a nook as this; the thickets of furze
are covered with golden bloom ; the young foliage of
the brambles is of the tenderest green ; clumps of
pale rimroses are carpeting the hollows, and fragrant
violets peep out beneath the bushes : the bee-orchis
is above ground by scores, and in the oozy outbreak-
ings of the springs rise hosts of the clubbed columnar
spikes of the great water horse-tail. The vivid-
coloured stonechat and the elegant rock-pipit are
flitting from hummock to boulder, selecting the scene
of their domestic economy ; and the perturbed haste
with which the blackbird shoots out from yonder
gorsebush, loudly cackling as he goes, indicates with
tolerable certainty his whereabouts. But here we are
at the beach, treading the leathery line that constitutes our hunting-field.

Many of these black and leathery sea-weeds at our
feet, which are shrivelling up in the sun and air, like
scorched parchment, are perfect microcosms. And
exquisite forms of being people these tiny worlds. I
pick out this broad frond from the dark mass, and
lo ! its expanse is studded over with tangled shrubs,
like the thickets of furze upon the downs yonder.
What can these little shrubs and bushes be ? " Sea-
weeds, of course," you say ;—" parasitic sea-weeds."

Not so fast, however. The difference between a

plant and an animal, down in these low regions of the organic scale, is not a matter of course; it is not a thing to be decided by a glance at the general physiognomy. These are *animals*, an't please you; truly sentient, conscious, wilful animals; as truly (I do not say as obviously) as that new-born lamb, whose caudal wriggle is the outward expression of inward rapture.

It is true that, on looking carefully over the patches, you discern no signs of animal life, nor of ordinary animal form. Take one of the least complex thickets. We trace the whole matted mass to a common origin —the springing of a slender stem from a number of diverging roots firmly adherent to the black surface of the Oarweed: this stem soon forks, and the branches so formed fork again and again, spreading themselves out, and crossing each other. These ramified stems are of a dull-drab or pale-brown hue, just like withered plants; and their whole extent is beset with tiny angular projections, like imperfectly-developed sheathing leaves, such as we see on many plants,—the broomrape, for example. The shrub has a creeping character, spreading over the surface, and here and there adhering, like a bed of verbenas that have been pegged down: and, as if to add to the resemblance and make the plant-like character perfect, you perceive, on looking carefully, that, at these points of contact, the branches have shot down rootlets, which have ramified and spread themselves over the surrounding area, adhering very firmly to the surface of the frond, which constitutes their soil.

With all these vegetative characters, it is hard to
believe that the Creeping Canda, for so it is designated by naturalists, can be anything else than a
plant. And yet a brief examination, combined with
a little knowledge, will suffice to convince you of its
animal nature; nay, that it is an animal not at the
very extremity of the scale, but with many a grade
interposed between it and the lowest. It is in fact
one of the same group of animals as the Tubulipora,
and belongs to the same grand division of animate
being as the Oyster, the Cowry, and the Snail.

I must tell you the life-history of this little shrub.
Some time ago,—how long I do not exactly know,
for we have not yet achieved any very reliable statis-
tics on the age and rate of growth of these creatures
—perhaps last summer,—an atom of living flesh,
scarcely discernible by the unassisted eye, even if you
had been present to watch it, might have been detected, by means of a lens, swimming in giddy circles
and rotating on its own axis in the open sea.   Under
the microscope, you would have found it a roundish
or pear-shaped animalcule, of a soft, yielding consistence, and therefore capable of changing its form at
will by irregular contraction. The whole surface was
beset with strong bristling cilia, or waving hairs,
which, acting as so many oars, rowed the little argo-
naut along on his circumnavigatory explorations.

After having pursued this course of wild freedom
for a while, it at length approached this broad frond
of Oarweed waving to and fro in the swell of the sea.

The animalcule settled down upon the smooth surface, but did not immediately affix itself, for it remained here gyrating on its axis, yet without leaving
the spot which it had selected, in the closest proximity to the leaf. At length, perhaps after an hour's
play, its rotation became momentarily interrupted,
and resumed with intervals of longer and longer duration, until at length it moved no more. The little
gemmule was permanently adherent.

Imagine the tiny, helpless atom of unresisting flesh,
now become immovably fixed to this exposed surface.
It has relinquished the only defence it possessed
against violence, that of swiftness. Its first need is
protection; and, Robinson-Crusoe-like, it immediately
sets about building a fortress. Nor is it, helpless as
it seems, at all destitute of resources; for, from the
whole surface of its soft body, ere now bristling with
cilia, a secretion is poured out, which, speedily hard-
ening, setting under water like the Aberthaw lime,
acquires a firm horny consistence, of considerable
elasticity, and perfect translucency. The softer parts,
which have hitherto presented a homogeneous substance without distinction of organs, now separate
themselves from the inner walls of the horny coat,
which thus becomes an enclosing case or cell; while
the flesh develops distinct and well-defined organs,
such as a crown of ciliated gill-threads around a
mouth, a long gullet leading to a stomach, an intestinal canal opening by a cloacal orifice, and many
threads and bands of muscular tissue, by means of

which the little creature can peep out of its fortress,
or retire into its shelter, or execute various movements
within its ample cavity.

If you had had an opportunity of examining the
Oarweed at that juncture, you would have discerned,
at the very spot where now you see the root of this
little shrub, the cell with its active inhabitant. You
would have seen an oval, or rather pear-shaped,
translucent shell, appearing as if a broad slice had
been cut away from one side, leaving a great aperture,
with a thickened edge all round. But this aperture
you would have seen to be closed by a delicate skin or
membrane stretched across it, connected with the rim
on all sides. Then, in the middle of this tense. membrane, or rather towards the upper extremity, you
would have noticed a kind of slit, forming two curved
lips, which closed of themselves when left alone, but
which constituted a door, capable of being pushed open
from within, and allowing the inhabitant a pretty free
egress. Beneath this membraneous coverlid, you
would have seen, in the interior of the cell, which,
from the perfect transparency of all the integuments,
would have been perfectly patent to your eye, the
little animal inhabitant, lying on its side, and bent on
itself with a double angle. You might have seen a
slight alteration in the position of the body, and then
you would have discerned the lips of the crescentic slit
in the membrane separating, as if pushed apart, and
a number of filaments, like a bundle of straight rods,
or like one of the fasces of a Roman lictor, slowly pro-

truding. Presently, these rods would simultaneously open out into a bell, or funnel-like form, and you would perceive each to be furnished with a double row of cilia, pursuing their ceaseless chase, exactly as in the *Tubulipora.*

After a time, this solitary animal began to increase; not, however, as yet, by a process of proper generation, viz., by the production of such a living, active, free-swimming germ, as that which originated its own life: this was an after-work, to be accomplished in good time. The first act was a multiplication of its own individuality, not the origination of a new race; and this was effected in a very plant-like manner. The primary cell had been strengthening its hold by shooting out, on all sides of its base, delicate tubular fibres of the common horny matter, which had crept along the surface of the frond before the substance had yet hardened, and had thus become so firmly adherent to it, that no force short of the actual destruction of the thread could have severed them. These creeping and branching threads had constituted so many holdfasts, which so far served the purpose of roots, though no nutriment appears to have been absorbed by them.

The anchorage being thus confirmed, a swelling began to appear from one side of the cell, which gradually developed itself into a bud, and finally into a second cell of exactly the same form, dimensions, and general character as the first. When completely formed, the two were soldered, as it were, together side by side, yet not exactly on the same level, the new

cell being placed considerably higher than the first.
It was occupied by a living tenant, in every respect
the counterpart of the former, with which it was connected by a thread of pulpy vascular flesh, which
passed through a perforation in the partition-wall of
the two cells.

From the angle formed by the second cell overtop-
ping the first, a third then budded in like manner, and
then a fourth, and so on; always maintaining the
same order, and thus growing in a two-rowed ribbon
of cells, each of which was intermediate in height between two others.

After the rising structure had proceeded in this
way for the length of perhaps half-a-dozen pairs of
cells, the angle was filled by a knob of the horny
matter, which did not develop into a cell, but budded
out a cell on each side of itself, in the same plane as
before, but divaricating at a wider angle. Each of
these cells now proceeded to increase in the regular
way; and the result of this was that two branches
were growing at once, the primary stem having forked.
And so the branches went on, growing and forking at
pretty regular intervals, until such a complex array of
spreading branches was accomplished as this which
we have in our hand.

We have already noticed the recumbent character
of the shrub, and the rooting of the prostrate branches.
This is effected by the shooting down of slender tubular fibres, which, on reaching the frond, divide into
a number of creeping radiating fibrils, of extreme

tenuity, which branch and re-branch, taking a firm
hold by adhesion of the surface, into which they have
no power of actually penetrating.

In order to see this with advantage, however, we
must take home the specimen, and submit it to a
good microscope. Then, by the aid of reflected light,
the appearance of the network of ramifying fibrils of
white shelly matter on the dark frond is very regular
and beautiful, and we perceive how effective a support is afforded by this contrivance. These little reticulate areas are thickly scattered on the leafy floor,
which is overshadowed by the umbrageous shrublet.

To the same revealing implement we must resort
for the full discovery of several interesting details
in the structure and economy of this minute creature. The habitation itself is far from being a simple
wall. Each pyriform cell is strengthened just above
its base by one or two thickened rings, forming a
sort of joint. It bears a certain resemblance to a
shoe, the large oval aperture representing the
" quarter," the edge of which, being thickened, is like
the binding of the shoe. It is a transparent glass
slipper, far more delicate than the foot of the fairest
Cinderella that ever lived could have put on. From
the outer edge of this " quarter," near the heel, project, in a slanting direction outwards, three short
shelly spines, the middle one being the largest, the
use of which, in the animal's economy, is doubtless to
ward off danger, whether arising from blundering
friends or malicious foes.

Besides these spines, another organ of defence is conspicuous, in the form of a branched projection guarding the membrane-covered aperture. From the inner margin of this aperture (or that which is next to the median line of the two-rowed branch) springs a shelly flat plate, which spreads laterally and divides into two, these into two more, and these in most instances with an incipient bifurcation again. Hence results a wide fan-shaped organ, which, arching over the membrane, protects it from violence, while it does not interfere with the projection of the animal out of the valvular opening, nor impede the ciliary currents which bring food and oxygen for its requirements.

Here and there, on looking carefully over several branches of the little shrub, we see one and another cell crowned by a globular body resembling a little pearl. This is an egg-cell; wherein are developed these gemmules, which when matured escape from their prison and swim away to form a new generation. These ova are not produced till the composite structure has attained considerable maturity; and the result of their liberation must not be confounded with that of the progressive budding forth of the compound tree. The one is analogous to the multiplication of a plant by cuttings, or rather by the successive growth of its twigs; the other to the formation of a new race of plants by the agency of ripened seeds.

Another interesting point, which would arrest your notice, is the number of long tubular filaments, which

straggle across the branches in various directions.
They are of great length, and are formed of the common horny substance of the cells, from which they
originate. If we trace them carefully, we shall find
that they always spring from a sort of socket placed
near the summit of the cell, on its outer side. Sometimes they seem to be branched, but most usually
they are obviously simple throughout; and as they
are connected by their extremities with other distant
branches, their nature is doubtless the same as that
of the root-threads, and their use that of affording
mutual support to the branches, which otherwise
might be liable to be broken off by the waves, or by
the trampling of crabs, and other animals.

By a little careful searching we may easily find, in
exactly similar situations, another species very closely
like this, but displaying some interesting points of
diversity. It is known as the Rugged Scuparia.
You would scarcely discern any difference between
this specimen and what we have just been examining,
except that this is rather more erect, less creeping in
its manner of growth. Nor, even when brought
under the microscope, would you instantly be struck
with the distinction. Yet distinction there is.

In this Scuparia the membranous aperture is not
protected by the arching fan-like operculum, nor by
any organ answering to it. But, as if to compensate
for this defect, there are some special organs of defence
which were wanting in the Canda. One of these is
called the *vibraculum*, or the whiplash. At the back

P

of each cell there is seated, at the point where it
springs from the summit of its predecessor, a kind of
oval knob, which is cleft to receive, as in a socket, a
more slender shelly knob. To this is attached, by a
very free hinge-joint, a long and very slender horny
bristle, tapering to a fine attenuated point. The
course of this lash is so curved, that it passes across
the front of its own cell. Its movements are curious:
ever and anon, fitfully and suddenly, it is swept forcibly along from one end of its range to the other.
It is believed that its use is to brush away any extraneous matters that, lodging upon the cell, might
interfere with the comfort of the inhabitant; or minute
intruders whose presence is felt to be annoying; or to
give a hint to larger visitors that their room is better
appreciated than their company. This conclusion is
strengthened from the circumstance that the range
covered by the vibraculum in its sweep, includes the
area of the aperture, where violence and annoyance
are more liable to be felt than in other parts.

Much more curious, however, than either operculum
or vibraculum, is the *avicularium*, or bird's-head.
The outer angle of the summit of each cell carries
what at first sight you would perhaps take for a stout
triangular knob; but when you look at it more carefully, you see that it is a very special and singular
organ. It is like the stout and strongly-hooked beak
of some strange bird of prey, cut out of its skull, and
soldered, upside down, to the angle of the cell. The
upper mandible (or what, in the true bird's beak,

would be the upper, though here reversed) is truncate, except the sharp-curved point, and encloses a capacious cavity; within this, the lower mandible, which is curved and pointed in like manner, is jointed, and, working on its hinge with an enormous range, shuts with a snap into the upper.

There are species (such as *Bicellaria ciliata*, *Acamarchis flabellata*, and others) in which the resemblance to the entire skull of a bird is most marked and striking. In these two the organ is not sessile and fixed, but attached by a hinge-joint, which permits great freedom of motion. A shelly knob is placed on the outside of the cell, and on this is seated the mimic skull, at that point where naturally it would rest on the atlas-joint of the vertebræ. The union is, as I have said, by a freely-working hinge-joint; and thus the whole skull sways backward and forward, just as a head does upon the spine.

But besides this, the form of the two mandibles is (in the former of the last-named two species, for instance) a far more perfect .copy of a vulture's beak. The *vraisemblance* is, indeed, most wonderful; and the microscope can scarcely present a more striking spectacle than one of these shrubby Polyzoa in full health and vigour in a trough of sea-water. The eye is bewildered and the mind amazed at the sight of scores of naked skulls swinging to and fro, not evenly and uniformly, but fitfully, and, as it appears, wilfully; while the yawning gape of the mandibles to an awful reach, and ever and again the spiteful snapping of the lower into the groove formed by the formidably-toothed edges of the upper, make us involuntarily shrink and shudder, lest the vicious bite should take a piece out of our flesh.

It is probable that this well-armed apparatus is auxiliary to the procuring of food. Not, however, that it directly subserves this end; for, being placed outside the cell, no communication exists between this ferocious mouth and the stomach of the animal within. But it may act as a trap to capture minute animals, which then, being held tenaciously until they decay, do in the process of decomposition attract millions of infusoria to prey upon them. These then, stimulated by abundant food, increase immensely in that vicinity by spontaneous fission: multitudes of these swarming minims must every instant be caught in the ciliary vortex of the expectant animal; and thus the action of the bird's head may be that of a man who over-night scatters ground-bait about the spot where on the morrow he intends to fish.

Babbicombe to Hope's Nose:

A MAY-MORNING WALK.

BABBICOMBE TO HOPE'S NOSE.

ANSTEY'S COVE, FROM BABBICOMBE DOWNS.

MINNICOMBE, Watcombe, Oddicombe, Babbicombe! "ferny combes" all. How characteristic of the coast of Devon is that suffix "combe!" To one familiar with the sweet shire it tells of a deep cleft in a precipitous sea-wall, a short abrupt valley, with or with-

out a tiny tinkling rivulet in the middle, with green
sloping sides sheeted with furze, bounded by tall cliffs
draped with ivy, and a shingle beach at the bottom.
And what a delight it is to roam along such an indented shore as this, when opening spring is just
clothing all nature with loveliness, at every turn getting a burst of some new secluded scene of beauty,
with the glorious sea ever bounding all !

Let us—you and me, gentle reader, *arcades ambo*—
set out together on such a walk; it will soothe our
spirits, quicken our pulses, heighten our joy, and,
maybe, deepen our gratitude to Him who " crowneth
the year with His goodness, whose paths drop fatness."

> " O evil day ! if I were sullen,
>   While the earth herself is adorning
>   This sweet May-morning;
> And the children are pulling,
>   On every side,
>   In a thousand valleys far and wide,
> Fresh flowers; while the sun shines warm,
> And the babe leaps up on his mother's arm ;—
>   I hear ! I hear ! with joy I hear !"

Along the margin of a cliff, now steep and sheer,
now breaking into an uneven but variously verdant
slope, we begin our march, ever and anon pausing to
gaze on the smiling scene below. The descent we
are just leaving behind, half-covered with the gorse
and guelder-rose, is Oddicombe, whose white crescent
beach lies below, bounded by the limestone promontory of Petit Tor, which divides the huge precipices
of red sandstone close at hand from the bluff coast of

the same formation that stretches away to the northward; its ruddy cliffs and bold headlands—Watcombe, the Ness at the mouth of the Teign, the perforated rocks and needles near Dawlish—gradually
fading into blue as the coast-line trends away to the
eastward, and is lost to the aching gaze somewhere
about the boundary of the county.

It is a lovely scene; and still more lovely is that
which meets the eye as we resume our walk and look
down upon Babbicombe beach, where fishermen are
overhauling their boats, already high and dry, and
the brown nets are spread out on the sunny shingle;
and where the whole slope is clothed with shrubberies
and hanging woods, with villas and ornate cottages
peeping from the embosoming trees, here and there,
down to the very water's edge.

And now it is optional with us, whether to pursue
our way along the seaward edge of the lofty Babbicombe Downs, or by the high road, which for nearly a
mile is shut out from the sea. We choose the latter,
as the more pleasing, and giving us more variety. It
leads us through the homely village of Babbicombe,
and then through the palatial domain of the Bishop
of Exeter, under the shadow of overarching elms; now
between hedgerows and banks bright with spring
flowers, now through deep scarps of the slaty rock,
dripping and ferny, and crossed with rustic bridges.

A burst of the sea again! Yonder it lies, sleeping
under the morning sun. Troubled are its slumbers
too: last night's easterly breeze has given it the night-

mare, for it heaves and tosses in its dreams like a free-
booter with murder on his conscience. 'Tis Anstey's
Cove that expands below us; another *combe* of most
romantic beauty, whose accessories are of more than
wonted grandeur, even on this magnificent coast. But
we shall have a better sight of it presently, and therefore we cross this rude stile on the left, and mount the
steep slope, threading a narrow path that winds through
the luxuriant furze. The eye is almost dazzled with
the golden radiance of its sheeted blossom, which is so
profuse and so gorgeous, that, if Linnæus could awake
again and behold it, he would go down on his knees
and worship it as he did of yore. The air, too, is redolent with its peculiar fragrance, and scores of aldermanic humble-bees are rifling its nectaries. But let
us walk cautiously; for on this hill the treacherous
adder lurks, and a careless foot is very apt to come
down upon the baleful serpent, as it suns itself on the
path, especially the gravid female, which, as if aware
of the advantage that the warm ray gives to the ma-
turing ova, is reluctant to move from the genial spot.
The early purple orchis is shooting up its beautiful
spikes of compact blossom in every nook; and, as we
rise to a higher elevation, other plants appear, so
minute as to be scarcely able to thrust their blossom
above the grass of the turf, short and low as this is.

I always like the tiny flowers of lofty downs; so
meek and unobtrusive, yet, withal, generally so pretty.
There are not many kinds yet abroad; but here is the
little dove's-foot cranebill, with its notched leaves and

sprawling crimson stems, and its pink flowers; and here the ubiquitous chickweed, with snowy stars. And what is this? Surely one of the bed-straws, as you may see by its spiny-edged leaves, set in many-rayed whorls, and by its fourfold blossom; but this latter, which is almost microscopically small, is of a decided lilac hue. And here, prettiest and tiniest of all, is the early scorpion-grass or hill forget-me-not, its slender stalks set with successive blossoms, all of which have the hues common to this bright-eyed and ever-welcome tribe, azure-blue with a yellow centre, except the terminal flower, and that is wholly yellow.

There is the whinchat! listen to his simple but sweet song; and yonder I see him perched on the uppermost twig of that furze-bush; the highest we can see. You may easily recognise him by his speckled back and wings of bright brown and black, his breast of bay, and the white band over his eye. He seems pouring out his soul in song; doubtless his nest is at the foot of that bush, or not far off; where, perhaps, his mate is cowering, listening to his music, and thinking it sweeter than that of all the nightingales in the world. Now, as if his spirit were too buoyant for his body, he springs into the air, and, hovering on expanded wings over the sacred spot, finishes his strain. Sweet bird, sing on! thou shalt not be molested by me.

Here we are at the summit, much to the relief of the aching muscles of our legs, and we stand at the very edge of a cliff certainly not much less than four

hundred feet high. A wide expansive prospect is on
every side. We will sit on one of these knobs of the
white limestone, that are everywhere cropping out
from the turf, all studded with incrusting lichens,
white, gray, black, and orange, and like swivel-guns
rotate on our pivots.

Look northward; down, down, the cliff-wall descends
perpendicularly for a hundred feet or more, then slopes
away, a wilderness of shrubbery, with great blocks of
gray rock projecting. Here, just before us, is a vast
buttress, upright, wall-sided, and round, like a battered
castle-tower of the olden time; and, like it, sheeted,
almost from base to battlement, with glossy ivy.
Ferns arch out from its crevices, and masses of the
curious navelwort, with its coin-like succulent leaves,
of unusual size.

Ha! like a stone from a sling, out shoots a large
bird from the rocky wall just beneath our feet; and
with a loud coo another quickly follows. Their form
and size, their manner of flight, and their colour, seen
clearly enough as we look down on their backs—blue,
with a conspicuous white rump, and black barred
wings—announce them to be rock-doves, the indubitable stock and original race of our domestic pigeons.
Away they go on loud whirring wing, and shoot across
the cove to the inaccessible ledges and clefts of yonder
precipice. It is by no means a common bird with us;
but a few pairs every season haunt the tall cliffs and
caves in this neighbourhood.

How noble is that huge promontory of many-tinted

limestone, projected like a Cyclopean mole into the
sea! Great stains of red, the washings of the red
earth above, itself the debris of more ancient red
sandstone, are seen on the rugged face of the cliff,
and at the bluff end, where the "jumper" and the
blast of the quarryman have done their work; having
followed the veins and cracks of the stone, and per-
colated through and through. Whymper has just
immortalised this magnificent promontory in one of
his beautiful water-colour paintings, and yonder is the
grassy cleft where I stood by his side as he sketched
it last autumn. It is curious to mark how large a
portion of the vast mass has been gradually quarried
away; for a long flat platform nearly level with the
water's edge, running out beyond the present sheer
end of the wall, and bearing two obelisks of stone
strangely preserved, shews the original termination.
The demand for limestone, some of which is a beau-
tifully-veined marble, for building purposes in the
vicinity, and for export, causes a constant diminution
of the mass; and vast as it is, the period is not at all
beyond contemplation when its magnificence will be
a matter of tradition.

The sea wildly dashes around the impregnable base
of the precipiece, and rushes with wild roar into the
dark caves, and makes mad efforts to scale the wall,
but always falls back in foaming rage, ever to essay
the assault again, and ever to be repelled by the passive resistance of the "everlasting hills." The pretty
secluded cove, with its white pebbly beach, offers a

no less effectual barrier to the breaking billows; they
run up, up, up, as if they would take the whole area
by storm, but are broken and dispersed, like a charge
of cavalry against a wall of British steel.

We lift our gaze to the summit of the great cliff.
It is almost as level as a wall, and crowned with a
thin stratum of short verdant turf, like that around
us. A single-coast guard is seen on the solitary
height, with a telescope—his invariable *fidus Achates*
—at his eye. He looms like a giant, as his dark form
is projected against the bright sky.

Turning toward the west, there is the episcopal
palace, an Italian villa, with its garden of terraces
and statues, and formal lines of cypresses, and par-
terres of brilliant colours; and the little old village of
Marychurch behind in the distance, loftily seated,
and its ancient square tower cleaving the sky. Farther to the south is Warberry Hill, whence a noble
panoramic view of Torquay, and much more, is commanded; the woods of Bishopstowe nearer at hand,
with a pretty new village church rising beyond them:
then rounded hills of turf, with a sweet little peep
of the sea lying in a cleft between, as in a goblet
half filled; a little glimpse of Torbay, blue and glittering, with many white-sailed craft speckling its
bosom.

Eastward lies the sea; the grim sea, the beautiful
sea, the many-sounding sea. Ships are swiftly scud-
ding over it, under a freshening easterly breeze, which
is covering it with " white horses," and breaking the

sunlight, that pours down upon it yonder, into ten thousand flashing gems.

But in this direction, or at least a little to the south of it, lies our further progress. Let us up and away, over these sloping fields, and through yonder coppice, and along the ridge which slants away to the shore, and ends in those ragged and bristling points of black rock.

Here we enter a quiet path, which is a favourite resort of mine. It is but a foot-track, winding through a thicket, or coppice, or hanging-wood—I scarcely know which to call it; it is all by turns;—almost immediately over a most wild and rocky shore. Sit for a moment on the step of the rustic stile! The mellow song of the blackbird comes up from the tangled bush below, so soft and sweet and rich; so flute-like, with a charming trill now and then; there is no rivalry; no answering note provokes him to emulation; his melody is soft and low, as if poured forth merely for his own gratification.

Hark! the cuckoo! O sweet cuckoo! O dear bird! thy two simple notes thrill my heart with a power far beyond that of the most perfect melody. It is the very breath of mature spring and early summer; the very expression of the loveliest season, when the year is in the very height of its beauty. Sweet cuckoo! thou hast given inspiration to poets age after age, from our early Gower down to Logan and Wordsworth. The quaint, but racy and forcible words of our earliest English poet come to my mind—

" Sumer is icumen in ;
    Lhude sing ' cuckoo ! '
Groweth sede, and bloweth mede,
    And spryngeth the wode nu.
     ' Cuckoo ! cuckoo ! '
    Ne swik thu naver ! "

An impudent magpie breaks the poetic spell with
his harsh cackle, and splutters out of that dark glade.
Magpie ! nay, magpies ! for there are two ! Of course
there are ; for who ever saw a magpie, without another
at his tail ? they always travel in pairs. Hech, sirs,
but maggy is a fine bird ! I never see him but I
fancy I see one of the splendid feathered denizens of
the tropics ; his bold contrasts of colour, his length of
tail, and its brilliant gloss of purple, green, and gold,
belong rather to the solemn forests on the Amazon,
or the sultry jungles of Borneo, than our chilly clime.
His voice though ! Well, that, I allow, is not melo-
dious ; but the parallel does not hold the less for that.

Just below us is a little grove of ash and stunted
oak ; and through the midst of this, which is dark
with the united foliage over head, a track leads
through the tangled thicket to the rocks beneath. It
is just passable, and that is all ; for everywhere it is
overrun with briar and bramble, and huge crowns of
the male-fern are crowded here in immense numbers
and prodigious luxuriance. It is quite a sight to see
their great fronds of filagree-work radiating and arching on every side. Moreover, as the blind path descends, it becomes more and more steep, and choked
up with loose blocks of stone ; until at length you

suddenly emerge on a great slippery rock, where there are only a few tufts of thrift to hold on by, and the beach yawning some thirty feet below.

But we are not going down to-day. And so, we saunter on our narrow path, now up, now down; now in the sun, now in the shade; now beneath an over-hanging block of cold rock, where water drips, and where the stonecrop and the navelwort grow, and the many-fingered polypody creeps about; and now under arches of foliage, a greenwood shade, where the sun-ray is reflected from a thousand dancing leaves. For the young trees are meeting and intertwining over our head; the hawthorn white with blossom, and filling the air with its fragrance; the sloe, the maple, and the guelder-rose with its snowballs, and the pointed heart-leaves of the bryony, so elegant in shape and so glossy in surface, hanging over every bush, as its long twining stems creep about like a network of living cords, a wild drapery of verdure.

The margins of our narrow footpath, too, are re-freshing to the eye. Coarse grass half hides the rough stone; the pale primrose is everywhere; the dog-violet, pretty, but, alas! inodorous, peeps up in companionship with it; thousands of white stars, like the constellations of a winter's night, mark where the stitchwort sprawls; the bright crimson blossom of the rose-campion, and the paler ones of the herb-Robert attract the eye; hundreds of the greenish-yellow umbels of the wood-spurge give a conspicuous character to the vegetation, and even the dog's-mer-

Q

cury aids the effect, with its light and feathery spikes.
But the hyacinth is the presiding *genius loci;* how
compactly do its smooth stems rise in serried rank,
each drooping with the weight of its numerous blue
bells ! All through the wood, in the tangled briery
thicket, and especially on each side of the path, as
far as the eye can see on either hand, there is a dense
belt of azure blossoms, reminding me of the ancient
Hebrew garment with its "fringes of blue" (Numb.
xv. 38). Butterflies are out, rejoicing in the advent
of spring: the garden-white flits to and fro amidst
the flowers; the speckled-wood dances up and down,
in its peculiar jerking way, over the herbage; and now
and then a tiny blue flashes out in mazy flight from
the groves, seen for a moment, like one of the
hyacinth blooms whisked about by an eddy of wind,
and then as suddenly lost to view.

How brilliant is the emerald hue of the young
foliage ! the limp and tender leaves of the beech, and
especially those of the ivy, shining as if varnished,
almost concealing in their profusion the old olive
foliage that bespreads the gray stone. See those
rugged masses, those huge angular blocks, those
peaks and obelisks, which, at some period or other in
the past ages, have been loosened by rain and frost,
and have plunged with mad crash and roar down the
slope, burying the shrubs and trees in crushed ruin,
till their own descending impulse was arrested. See
how Nature is ever asserting its restorative power !
Naturam expelles furcâ, tamen usque recurret. A

season or two conceals the damage, and a few years repair it; and then the shrubs and the briers grow up around the hard-featured intruders, and the creepers and ivy embrace them, and gradually envelop their rugged sides and angles, till the intertwining tendrils meet above, and the evergreen drapery presents a continuous surface as before. Already the work is half achieved; and by and by not a sign will be seen externally of what seemed at the time horrid and incurable wounds.

But there is a line where the vegetation ends, and gazing down from this steep, the eye at length comes to a broad belt of ruined rocks; fragments wild and ghastly, of all rude shapes and of all dimensions. These, too, have been dislodged from above, and, unlike their more favoured fellows, have plunged beyond the region of shrub and brier, ploughing their wild way through all, and there they lie in chaotic confusion, heaped on one another, without a leaf or blade of verdure to break the bald blackness of that broad belt of ruined boulders. It is truly a "line of confusion, and stones of emptiness ! "

The tide is in. The encroaching sea insinuates itself among the masses, rising and falling, and seething up through the crevices, and closing over the broad surfaces, the next moment to fall in green cascades over every side, or covering the black rock with sheets of pale blue and white foam, and tossing around wreaths of feathery spray over the peaks, or shooting up through some narrow crevice a tall jet

of water, with a sucking sound and a report like that
of a rifle; while, without intermission, there goes on
that melancholy wailing, washing hiss, which is the
constant accompaniment of a breaking surf.

And now we have reached the end of our pleasant
path. The luxuriant verdure ceases; there is no more
shrubbery or coppice; but one or two fields belonging
to Ilsam Farm are laid down in grass, and beyond
these there is a gentle declivity of down, with its
clumps of gorse; and beyond that a low, broad promontory of naked rock.

This is Hope's Nose: the northern propyleon of
that indent of the coast called Torbay, as bluff Berry
Head is the southern one. Three rock islets lie
around this point, like sleeping lions guarding the
gate; and on one of them there is that interesting
geological phenomenon—a raised beach.

What a picture of utter desolation is presented by
this promontory! It reminds me of what travellers
tell concerning Sinai and Horeb, and the land of Edom,
—*magna componere parvis*,—a great area of weather-
stained limestone, split with fissures in various directions, and most strongly contrasting with the soft,
wild, luxuriant beauty, amidst which we have just
been rambling. To add to the ruin, man has been
here, too, quarrying; and the great rugged excava-
tions, all coarse and angular, and the heaps of rub-
bishy *débris* at foot, make it a miserable place to look
at. Let us hasten on to the extreme point.

Here is a flat platform of the same gray stone, compact and solid in its own substance, but much shattered and split. Deep narrow clefts, with wall sides, penetrate far in, into which you can look down and behold the sea raging. If it were low water and calm, you would see a splendid sight in these fissures; for their perpendicular sides are studded below tide-marks with various species of anemones, alcyoniums, and other zoophytes, by no means of common occurrence, in amazing profusion. But the sea penetrates much farther than you would suppose on a cursory glance. Look down any of the irregular crevices, and you will see the sea at the bottom, and by peering obliquely into the fissures, you will perceive that the whole of this great platform is undermined, and actually overhangs the sea.

Other evidence of the same fact forces itself upon us in a somewhat unpleasant manner. The muffled roar of the billows is heard beneath our feet; and at every wave a blow is given to the solid stone on which we are treading, the shock of which is distinctly felt, imparting a peculiar nervous sensation, perhaps not unakin to that produced by an earthquake. We scarcely like to stand here; though the permanency of the area from year to year tells us that there are pillars stout enough beneath to assure our safety. However, we will be going.

Before we leave, I will just indicate the situation of a little rock-pool of peculiar luxuriance, a thorough little tank of marine zoology, a well-stocked aquarium of beauties. It is readily accessible, being placed at

the very margin of the extreme point; but so over-
arched with a projection of the rock, and so concealed
by oar-weed, that it would be very likely to escape detection, unless previous knowledge pointed it out, or
accident revealed it. Here it is; but as we are not
out anemone-hunting to-day, we will for the present
postpone a minute examination. And now we will
retrace our steps, musing on what we have seen, and
on the love which has made everything so full of
beauty. Every scene reflects His glory, every sound
is vocal with His praise. Happy the man of whom
it can be truly said, in the words of our own sweet
Cowper:—

> " He looks abroad into the varied field
>   Of nature, and though poor, perhaps, compared
>   With those whose mansions glitter in his sight,
>   Calls the delightful scenery *all his own.*
>   His are the mountains, and the valleys his,
>   And the resplendent rivers. His to enjoy
>   With a propriety that none can feel,
>   But who with filial confidence inspired,
>   Can lift to heaven an unpresumptuous eye,
>   And smiling say—My Father made them all !
>   Are they not his by a peculiar right,
>   And by an emphasis of interest his,
>   Whose eye they fill with tears of holy joy,
>   Whose heart with praise, and whose exalted mind
>   With worthy thoughts of that unwearied love,
>   That plann'd, and built, and still upholds a world,
>   So clothed with beauty, for rebellious man ? "

AN HOUR AMONG

The Torbay Sponges.

AN HOUR AMONG

THE TORBAY SPONGES.

TORBAY : BERRY HEAD IN THE DISTANCE.

A FEW days ago, I made my first regular hunting-day of the season among the sea-creatures. Those who are given to the study of out-door natural history, in any of its numerous branches, will know the delight with which, on a lovely, balmy, sunny morn-

ing in April, one calmly lays aside study, correspond-
ence, work of all sorts, and resolutely says, " Stay you
till to-morrow; to-day I go hunting !" Winter is
over and gone; at least we persuade ourselves that it
is : the day has opened in cloudless glory. " Will it
last ?" some one endowed with the bump of cautious-
ness asks. " Of course it will, for are not we going
anemone-hunting ?" However, to make all sure, we
can put umbrellas and shawls and cloaks into the
carriage. The worms and molluscs will have come
into the shallows by this time, after the winter, for
the depositing of their spawn; and will be sure to be
found under the stones and in the crannies. The
tide, too, will be of unusual excellence; it is the full
moon after the equinox, the very best spring-tide of
the half-year. We may expect an immense reach of
coast to be laid bare soon after the sun begins to
decline from the meridian. The wind is off-shore,
and has been so for some time; so that there will be
no sea running, and we may explore to the very verge
of low-water. Everything is propitious : why do we
tarry ? We do not tarry, for the carriage is ready,
and we bundle in, the whole household, all intent on
a day's hilarity ;—

> " All agog,
> To dash through thick and thin."

But what need of a carriage, seeing I reside at
Marychurch, with a capital shore, varied with cove
and headland and cliff, with sand and shingle and
boulders and rocky ledges all round me, approached at

several points by half-an-hour's easy walk? Ah!
gentle reader, I'll whisper a secret in your ear; but
don't tell that I said so, for 'tis high treason against
the ladies. Since the opening of sea-science to the
million, such has been the invasion of the shore by
crinoline and collecting jars, that you may search all
the likely and promising rocks within reach of Torquay, which a few years ago were like gardens with
full-blossomed anemones and antheas, and come
home with an empty jar and an aching heart, all
being now swept as clean as the palm of your hand!
Yet let me do the fair students and their officious
beaux justice: the work is not altogether done by
such hands as theirs; but there is a host of professional collectors, small tradesmen whom you must
search-up in back alleys, and whose houses you will
easily recognize by the sea-weedy odour, even before
you see the array of pans and dishes in front of the
door all crowded with full-blown specimens. These
collect for the trade, and are indefatigable. Only
think of the effect produced on the marine population
by three or four men in a town, one of whom will
take ten dozen anemones in a single tide!

The fact is, the fashionable watering-places on our
south and west coasts are completely stripped; and
any one who really wishes to find anything worth
having, must seek some quiet, undisturbed sea-nook,
where there are no visitors, where the new trade has
not yet been set up, and where the poor people are
too primitive to notice such " rubbish" as you value.

Therefore it was that we ran some miles away
from home, and pursued a pleasant road, partly
through green lanes, rank with the glossy young
leaves of the arum, and the arching fronds of the
hart's-tongue fern, scarcely embrowned by the late
arctic winter; and partly sweeping along the shore-
line and over the cliffs that make the base of this
beautiful bay; till, Paignton being some distance
behind us, we turned off to the left down a little lane,
and drew up at the margin of the broad flat beach
called the Goodrington Sands.

Far away is the edge of the sea, for the tide is
wonderfully low, though we have yet a full hour and
a half before it will be at its lowest point, and an
immense breadth of soft, wet sand lies exposed. We
pause for a moment to gaze on the boundary to the
right.   It is Berry Head, a noble headland that
projects like a long wall far out into the sea, and
presents its bluff termination, crowned with fortifica-
tions, to the impact of the waves that drive in with
impotent fury from the wide Atlantic.

But now to work.   Out with the collecting baskets,
the bottles and jars, the stout hammer and the strong
steeled chisel, and away across the heavy sands, in
which we sink at every step, away obliquely to the
left, where another bold headland, Roundham Head,
breaks the sweep of the bay, and for the present shuts
out Torquay from our view.

There is our working ground, at the foot of those
red cliffs. We diverge a little from a straight line, and

approach the edge of the sands, in order to see what those two men are so busy about, as they trudge along the water-line with stooping backs and downward gaze. Oh! they are fishermen taking solens, or razor-fish, as they call them. Each carries a light, narrow, but deep spade in his hand, and, as he marks a little jet of clear water that spirts upward from a small hole in the sand, he rapidly thrusts in his instrument, and adroitly jerks out his prey. It is that mollusc, whose long parallel-sided, convex, bivalve shell, something like the handle of an old-fashioned razor, is so common on every sandy beach, but which is more rarely seen alive. Here we see the poor creature so unceremoniously brought to light, much too big for its valves to contain, its pellucid body shrinking and quivering, its long white foot, like a finger cut off slantwise, and its siphons still contracting, and discharging the limpid water in great rapidly successive drops. The man scarcely deigns it a glance, thinks nought of its curious structure, cares only for the halfpence it will bring him in the fish-market, jerks it into his basket, and watches for the next jet of water with which the frightened and retiring mollusc shall betray its place of retreat.

We quicken our steps to atone for this momentary delay, for time is precious, and the tide has not long to run; and time and tide wait for no man. Now we approach the wilderness of boulders that fringe the cliff-foot, huge masses of the coarse red conglomerate, that the combined action of successive winters and summers has dislodged from the promontory, and plunged in confusion at its base. The unwonted recess of the water to-day permits us to wind around the outer edge of these, with a little shallow wading, and an occasional climbing over a block more obtrusive than its fellows. We soon see that we are on promising ground. The perpendicular surfaces of these huge blocks, especially those which are turned from the sun, are crowded with specimens of various species of anemones. Here are numerous colonies of the smooth, great, overgrown strawberries, displaying their yellow-green spotting on their liver-red bodies; smaller, but more attractive self-coloured ones, plump and pellucid, crimson and green, reminding us of cherries and green-gages; and hosts of little ones, hardly arrived at an age to develop any particular character as yet. And here are great daisies, profusely crowded; purple-bodied, pink-based fellows, lolling out of holes and crevices in the coarse rock, and inviting the honour of capture. We cannot resist the temptation; we apply the chisel, and by a few well-directed strokes of the hammer, succeed in separating large pieces of the soft and friable stone, which, loaded with the uninjured daisies, are thrown into the basket loose; the glass jars being reserved for tenderer and more delicate things.

Many of these blocks of stone, having fallen one upon another, are supported in such a manner as to make arches and low-roofed passages; and such conditions are sure to be prolific in marine life. We

stoop down to peep beneath them, and see the under
sides of these suspended ledges swarming with strange
forms of many-coloured existence. Slimy and wet,
indeed, they are, and coated with an impalpable mud,
the deposit from thousands of molluscan stomachs,
which are continually rejecting the indigestible portions of their food in this form; and hence it is impossible to explore such situations without horrid
defilement of the garments. However, this is no
great evil, for we are accoutred for the occasion; so,
kneeling, or semi-reclining on sides and elbows, or
fairly stretched at full-length supine to give the
hands full play, we worm ourselves into the holes
and crannies, and gather till the jars cry " Hold !
enough ! "

Sponges of various forms and hues delight in these
situations. Some, of a yellowish brown colour, form
large patches, everywhere throwing up little perforated
cones, like the hills in a region actively volcanic;
others of nearly the same form, but of a lovely pellucid
rose-colour; others forming a spreading leprous crust
of the richest scarlet, periorate, but not forming
conical eminences. Others again, of rarer forms,
branch out in ramose shapes, tubular and trumpet-
like, of pale yellow hue; and some, stiff and cartilaginous in texture, form regular tumours of dirty white,
ribbed and scored, standing up from the surface of
the rock. Others again form a congeries of little pipes
of snowy whiteness, exceedingly delicate, and ramifying and again uniting at all angles; and others take

the shape of flattened sacks, attached by a base to the
red leaves of depending sea-weeds.

All these are Sponges. Diverse as they are in form,
and in texture, and in colour, and in manner of growth,
they have all the same essential structure. We cannot learn much about them by looking at them here,
especially after they have been for an hour or more
forsaken by the receding tide; but if we take one or
two specimens off very carefully, so as not in any wise
to bruise or break their delicate organization, separating, in short, by means of the chisel, a bit of the rock
itself on which they are growing, and, committing
them to a jar of sea-water, examine them at home,
we shall find much to admire in these, the lowest
works of God's hand, and see abundant occasion to
praise His infinite wisdom and inexhaustible resources,
and to render to Him, what the study of creation
ought ever to elicit, the glorification of His eternal
power and Godhead.

Let me then intermit, for a few moments, the
description of our shore explorings, and tell my young
readers somewhat of the wonders which these Sponges
reveal when examined under favourable conditions
at home. We will assume, then, that the specimens
have been safely brought home; that they have been
lifted one by one from the collecting jar, and, with a
soft camel's-hair pencil, have been cleansed from extraneous matter, mud and other deposits, while under
water; that each has then been rinsed and deposited
in a small glass cell with parallel sides, full of fresh

and clean sea-water, and left for twelve hours at least. Then, taking care not to touch the glass cell, nor to jar even the table on which it is placed, either of which might cause the sensitive sponge instantly to cease its operations, we bring a powerful pocket-lens close to the glass, and intently watch the specimen within. Suppose it is one of the yellow species, which throws up little hillocks, the Crumb-of-bread Sponge; our attention is at once excited by seeing a strong movement in the water, through which tiny atoms are hurried along in swift currents. We fix our gaze on one of the hillocks:—lo! it is a volcano indeed! From the perforate summit of the cone, as from an active crater, is vomited forth a strong and continuous stream of water, and crowds of atoms come pouring forth, disgorged in succession from the interior, and projected far away into the free water, to be followed by unintermitting crowds of others. This is highly curious, and we wonder what is the nature of the power which so strongly conveys to us the idea of an active vitality in a mass so inert and apparently life-less as this yellow encrusting sponge.

But let us apply the magnifying glass to another cell containing one of the bits of rock that has on it the cones of pellucid rosy lilac, the Rosy Crumb Sponge. There is a general resemblance to the for-mer in shape, as it is an encrusting kind, spreading over the stone, and rising here and there into well-marked conical eminences, each of which is perforated with a large circular orifice. The colour somewhat

R

varies, for while some specimens are of a fine red-
purple hue, others are lilac, and others fading to almost white. As we look at our specimen through the
magnifying lens, we fancy that the eye roams over an
undulating country studded with pointed hills. But
a peculiarity which at once strikes the observer, and
to which there was nothing parallel in the former case,
is, that the whole surface of this mountainous region
is studded with tall and slender poles projecting from
the ground, at various angles with the horizon, and
frequently set in little groups. These poles or rods
are drawn to a point, transparent, and seem made of
glass.

More unaccountable still, we see a web of exceeding delicacy, far more delicate than the finest cambric,
transparent and colourless, thrown over the entire
hilly region. It appears to have been spread after the
rods had been inserted, or else these have protruded
from below, under the investing web ; for though here
and there the points have pierced through it, yet they
have lifted it from the surface, and carried it partially
with them, so that it hangs in crescentic veils from
group to group.

Is the region we are looking at, then, sown with
precious seeds, or bearing some very rare and valued
fruit ? And is this a web of netting thrown over the
whole, and supported by poles, in order to protect the
crop from the ravages of greedy and intrusive birds ?
Not at all. Take away the magnifying lens, and the
object in an instant shrinks to its true dimensions and

its natural character : it is but a tiny bit of mammillary sponge, some inch and a half in diameter. Let us then drop our comparison, and think of it as— what it is.

Here, as in the former case, a strong intestine motion is visible in the water to the unassisted eye ; and the lens quickly enables us to trace this to a powerful and constant ejection from the mammillary orifices. But it is far more forcible in this example than in the other. In order to see it to advantage, we must rub a little carmine on a palette, and with a camel's-hair pencil diffuse carefully a small portion of the fluid pigment in the water of the cell, making it slightly dimmed with pink clouds. Then lifting the cell upon the stage of the microscope, but so cautiously as not to give the least jar or shock to the contents, we must apply a somewhat low power, about seventy or eighty diameters, for instance, and anew watch the result.

It is beautiful now to see—the process having been performed so gently that the living action of the Sponge has not for a moment been intermitted,—how the water, loaded with the atoms of red pigment, into which the magnifying power is sufficient to resolve the clouds, is uniformly drawn from all surrounding parts within a certain range towards each orifice, slowly and imperceptibly in the remoter parts of the circle, but ever acquiring more and more velocity, till it rolls up the sides of the hill, and then is shot away perpendicularly like a torrent of smoke and ashes

from the crater of a vomiting volcano. The object of
the infused pigment is, by filling the water with
opaque atoms, without destroying its fluidity, to render the motion of the currents much more appreciable
oy the eye.

There is, however, another interesting phenomenon
to be exhibited by our little Roseate Sponge, but this
will be manifest only after a somewhat lengthened
period of undisturbed rest, and in a larger volume of
water than the stage-cell will contain. Removing
the specimen to a straight-sided tank, and placing it
so as to be very close to the glass wall, we wait a day
or two, and then bring the microscope (which must,
in this case, be of that construction which allows the
mirror and stage to be removed, and the body to be
screwed to the edge of the table) horizontally opposite
the tank-side, a window (or a lamp, if at night)
being on the farther side. Now we see with delight
that from the interior of the mimic crater there is
projected a membranous tube, of the most exquisite
translucency and delicacy of structure, and extending
to such a height as to be distinctly visible to the naked
eye. Its lower extremity is commensurate with the
aperture from which it issues, and the upper contracts
to a narrow circular orifice, about half as wide as the
tube. It is apparent that this projecting membrane
is only a continuation of the common web-like tissue
that invests the entire surface, and it bears, imbedded
in its substance, especially at its lower part, some of
the glassy rods, yet not so thickly as to interfere with

its general clearness. If touched, it does not shrink at once, but if we remove the specimen from the water, and presently replace it, we find the tube so shrivelled as to be invisible; though by patiently watching for a few minutes, we perceive it slowly re-appearing, minute at first, and closed at the extremity, but soon acquiring its former dimensions, and gradually opening its terminal orifice.

We have, however, found other kinds of Sponge besides these at our rocky point. Now it frequently happens in natural history that, though endowments, and faculties, and properties are common to several allied species, some one of these is observed to most advantage in one species, and some in another. This projection of delicate gelatinous tubes, a highly curious and interesting phenomenon to witness, for example, is better shown by the low vermillion crust of the Sanguine Sponge, than by the mammillary hillocks of the Roseate sort. This is a very common species, and one which, by its scattered patches of brilliant scarlet, much assists to give that rich variety of colour which our rocks display when exposed at the lowest spring-tides. Examining, then, our specimen of this beautiful Sponge, with the same appliances and the same care as we used for the others, we obtain the following results.

I may compare the Sanguine Sponge to an *uneven*, rather than a *hilly* country, the eminences being uniformly lower, and very irregular in shape and elevation. Perforations appear here and there, like deep round

pits sunk in the soil; and, as before, a transparent
web is spread over the whole, which is forced up by
projecting glass rods; these are not pointed, but blunt,
as if abruptly cut off.

From one and another of the pits, a round bladder
is seen pushing out, which gradually lengthens till
it becomes elliptical. It is formed of a film of the
clearest jelly, excessively subtile, yet tenacious, with a
yellowish coat of granules spreading irregularly over
it. Orifices are now seen forming in the rounded tip
of the bladder; the origination and enlargement of
which are so very gradual as to defy detection, except
by marking stages of progress. The number of these
orifices varies up to half-a-dozen, as do their size and
position, both which are quite irregular : they are,
however, invariably situate at the extremity of the
bladder. They have always a well-defined outline,
which is rounded in every part, except where two
contiguous ones are divided by a slender thread of the
common membrane, in which case the two form the
halves of a circular or ovate figure. Slowly and imperceptibly they are seen to change their size, augmenting or diminishing; sometimes a minute orifice
appears at the margin of a large one, and increases at
the expense of the older, until the dividing film is reduced to a thread stretching across, like a narrow
causeway across a lake. Now and then, from some
cause inappreciable by the observer, the whole bladder
will wrinkle, and partially collapse into a slender

roughened column, then slowly distend again, when the orifices are seen to have not changed.

It is at these orifices of the protruded bladder that the current from within is poured forth, carrying with it the fœcal residua of the assimilated particles, and any light floating atoms that may be in the vicinity, as we see with admirable distinctness, by mingling with the water a little paint or a minute quantity of powdered chalk. It is then very interesting to trace the particles that are vomited forth, all up the interior of the bladder — its perfect transparency allowing them to be clearly discernible—from their first issue out of the pit, till, rolling over and over in their course, they are shot out of the extremity of the bladder, and involved in the ever-mingling currents.

The bladder in this case manifests much more sensibility than that in the Roseate kind; for, on being touched with the point of a needle, it immediately wrinkles and shrinks up, though it does not retract itself, and presently dilates and distends again. It is, therefore, not true that the Sponges manifest no movement or contraction under the impact of extraneous bodies, which might indicate the slightest perception of touch. The spongy mass, indeed, does not shrink, and probably this has been generally the only part tested, the extremely delicate protusile membrane having been overlooked. With such experiments as I am describing, however, Sponges will exhibit with great distinctness two characteristics of

animal life,—spontaneous movement of parts, and
sensibility to touch.

The thoughtful observer, watching the evolution of
this unintermitted current, ever pouring out with such
power and velocity and volume, would ask, What is
the nature of the force that vomits forth the fluid?
what its seat? and whence the supply? No visible
current passes inward from without; still, as the
stream is continuous, and yet the quantity of water in
the cell does not increase, it is manifest that the water
from without must enter in the very same ratio as it
is expelled. In order to understand this, we must
cut or tear a Sponge to pieces. We shall find that
the round apertures are the mouths of a few large
canals which run through the interior; that into
these open, at irregular intervals, other subordinate
canals; that these receive others smaller still; these,
again, others, in an ever-diminishing ratio; till at
last we can no longer trace them as canals, the whole
superficial portion of the Sponge being pierced with
microscopically minute and innumerable pores. Into
these the external water is constantly being absorbed,
carrying with it both oxygen for respiration, and
organic matter for nutrition. The influent water,
parting with these elements, and thus revivifying the
living gelatinous flesh that clothes every fibre, gradually permeates the whole interior, flowing along
the pipes in succession, till at length it gathers into
the larger canals, and is poured out at their apertures,
as we have seen; just as the waste water from every

house of a large city falls down the sinks, and rolls through the smaller sewers till it reaches the main, and joins that of other houses, and is vomited forth at the common outlet. Or like the rains and dews, which, falling noiselessly and unobtrusively over a great extent of country, collect in mountain springs, which feed the rivulets and brooks, and these in their turn unite into rivers, which open on the coast in a broad estuary, and send forth a volume of fresh water, whose current can be perceived for many miles in the open sea.

A Sponge is composed of a clear granular jelly, investing a fibrous or spicular skeleton, formed of horny matter, or flint, or lime. The Sponge which we use for washing has a skeleton made up of fibres of horn, but those which I have been describing have their solid parts made up of flint, the particles of which are arranged in needles (*spicula*) of a perfectly transparent, solid, brittle glass. These are the rods, a few of which we see projecting from the surface in the Roseate and Sanguine Sponges, but which, on examining a minute atom of their substance under a high microscopic power, we find to be incalculably numerous in the interior. The gelatinous flesh has the power of secreting the flint from the sea-water, and of depositing it in regular needle-like forms, and in such an arrangement as to produce the canals and apertures that I have described above. The flesh itself is furnished, on the surface that lines the canals, with curious filaments or hairs called *cilia*, which are

endowed with the faculty of waving to and fro in
given directions at the will of the animal (for, strange
as it may sound to some of my readers, a Sponge is,
beyond all controversy, an animal), and in rhythm
or harmony with one another; and these regular
wavings impart movement to the water, and cause
currents to flow in a given direction through the
canals.

These *spicula* or needles, however, that make up
the firm portion of the Sponge, are worthy of a little
notice. Without them the creature would be a mere
drop of glaire, having neither form nor consistence.
And yet a heap of needles seems to have little power
of assuming or of keeping any definite corporate form,
when we remember that they have no adhesion to
each other, and nothing, in fact, to keep them together but their mutual interlacement, and the thin
glaire by which they are invested.

In order to obtain a good idea of their structure
and appearance, we should take one of the white
species, such as the flattened sac that is so commonly
seated on the stems of the red algæ, several of which
we have taken at Roundham Head.

In these the constituent substance of the *spicula* is
not flint, but lime; but the delicacy, beauty, lustre,
transparency, brittleness, smoothness, and fineness of
the glass are the same in both cases. As we must
use a very high magnifying power, so we can observe
only a very minute portion at once; hence the best
mode is to tear one of the little sacs apart, and with

a needle-point separate a bit the size of a rape-seed, and, laying it on a slide of glass, let fall on it a drop of concentrated potass. The fleshy gelatinous invest-ment will thus be dissolved, and the spicula will appear alone, but undisturbed in arrangement.

What a wilderness of brilliant starry points now meets the eye! An incalculable multitude of three-rayed stars is seen, as if three needles of glass had been united by their heads, so as to radiate at an angle of 120 degrees. There is no variation in the angle of radiation; all are exactly alike in this re-spect, though they differ much in the length and stoutness of the rays. They seem as if thousands upon thousands of such stars had been put into a box, and well shaken together, so as to be inextricably interlaced. Some seem, naturally enough, to have been injured by the shaking; for many a point is broken short off, at varying distance from the diverg-ing centre. Of course, this shaking together is only imaginary; only a homely comparison to help de-scription; the real explanation of the entanglement doubtless is, that they have been deposited by the living flesh, atom after atom, in succession, and that the points of the newly-formed have shot between and among the interstices of the already existing ones, producing such a tangle that it would probably be impossible, even with pliers ever so fine, to extract one from the mass, without breaking either itself or some of its fellows.

Of course, I found many more objects of interest

in my couple of hours' explorations at low tide among
yonder rocks, and some of these I may speak of here-
after. I hope some of my readers may be interested
in these attempts to describe atoms that are among
the meanest things which God has made. I say " attempts to describe" rather than "descriptions;" for as
I gaze at the wondrous array of starry spicula actually
spread out on my microscopic stage, on the table at
which I am writing this paper, at this moment, I feel
how inadequate are words to grasp the inconceivable
perfection and glory of the Divine handiwork. It
matters not what the structure be: it may be the
bony casket that shields the brain of man; it may
be the cells that make up the petals of a painted
flower; it may be the needles of a sponge picked
from the mud of a tide-forsaken rock; the inimitable,
unapproachable, incomprehensible impress of Deity is
there. Augustine says, " The soul bending over the
things Thou hast made, and passing on to Thee who
hast made them, there finds its refreshment and true
strength."

Thus would I desire to contemplate the works of
God, as bringing to my sense ever-fresh proofs of His
all-pervading care, of His wondrous skill and wisdom,
of His glorious majesty and power. Above all, they
are the productions of the august Word: it is not
that they were made by One who is infinitely great,
but far removed from me, so that I can only rever-
ently admire Him at an immeasurable distance. No;
they are the productions of the mind and hand of the

Word (John i. 3); of Him who in His unfathomable love came down and took hold of my nature; took hold of *me* (oh blissful thought !), of *me*, a lost, guilty, ruined creature, sinking into inevitable perdition. He humbled Himself, emptied Himself, took my guilt upon Himself, united Himself with me, and me with Him, till, triumphing over ruin and death, He bore me up with Him, in indissoluble union, to His own seat at the right hand of the Majesty on high (Eph. iii. 5, 6).

Yet let me not be mistaken. The study of the creatures could never teach me this. Notwithstanding all that they eloquently declare of the eternal power and Godhead of the Creator, they are ominously mute when I ask them how He will deal with me, *a sinner*. I see, indeed, His boundless goodness; I see with admiring wonder the contrivances and arrangements put in motion for the health and welfare of a poor sponge. Surely this tells us God is good ! Yes, yes; no doubt of it. I see He is good to the sponge; but the sponge is not a sinner, a rebel, a contemner of Him,—all which I am. How, then, will He deal with *me ?* Ah ! nature gives me no light here; not a ray to brighten my darkness. It is to the written Word; it is to the Book, that I turn, and there I learn the mystery of the incarnation; the redemption of a sinner by blood; the payment of my infinite debt; my union with the God-man in resurrection life, on my believing the record; and the sharing of His coming glory.

Goby Hunting.

GOBY HUNTING.

THE GOBY IN THE TANK.

ONE of the most pleasant, because most certainly suc-
cessful, modes of prosecuting the natural history of
the sea-shore, is what is known to the initiated as
" stone-turning." We look out for a beach where
large stones—masses, I mean, which we are just able
to turn over with ease, lie pretty thickly strewn;
flat irregular-shaped masses, well clothed above with

s

GOBY HUNTING.

a dense forest of sea-weed, shewing that they have
been undisturbed for some months at least. There
should be little else than such blocks as these, lying
one on another, so as to leave clear the narrow cavities beneath, through which the sea may play; for
stones, however inviting they may look on their upper
surface, are next to worthless if they are imbedded in
sand or fine shingle. It is not by any means *every*
beach that yields the necessary conditions; some
beaches are all sand; these have their own proper
creatures; a sand-beach is not at all barren of animal
life, but it does not give us what the stones yield.
Others are all shingle, made up of smooth, sea-
washed, rounded pebbles, from the size of a marble
to that of a turkey's egg, which ever roll over one
another, with a whispering sound, as the surf runs
in: these are utterly barren; the very worst of all
localities for the marine naturalist to try his fortune;
the small size of the stones renders them so movable
that nothing can adhere to them with permanence,
and they afford, for the same reason, no available
shelter for darkness-loving creatures! A mixture of
large stones with shingle is but little better; for the
pebbles wash in and out between the heavy stones,
and not only fill up the interstices, but, by rolling
and rubbing, effectually clear their surfaces of all
adventurous atoms, animal or vegetable, which might
essay to take up a residence upon them. Often, however, a beach, which presents nothing but unmixed
shingle from below half-tide level to highwater mark,

becomes changed in character from the former range
downward, and, at the level of spring-tide low water,
is wholly composed of promising stones. The prac-
tised eye soon recognises the suitable spot, and is able
even to distinguish among the stones themselves those
which—from something in their appearance, which
perhaps could scarcely be intelligibly described; something in their form, or texture, or position, the manner
in which they lie, and their relation to the surrounding stones—are most likely to reveal hidden treasures.

We have found such a beach (I know of a very
good one not far away; all the better, because the
amateurs and the trading collectors do not much in-
vade it), and begin our examination, about an hour
before the ebbing water reaches its extreme point,
on a good spring-tide, the second or third day after
full moon, we will say, in April or October, for then
the tidal wave recedes the lowest. The wind is off-
shore, and not a ripple is breaking the mirrory sea;
there is no swell, the remnant of a storm now hushed
to repose, so that we may work at the very lowest
verge; the stalwart young fisher returning from
examining his nets has hauled up his boat upon the
beach not far from the spot we have selected as the
scene of operations, and in his high water-boots, striped
Guernsey shirt, and red liberty-cap, lolls against her
bows, and, as he puffs his short cutty-pipe, looks
askance at us, half curious to see what we catch, yet
disdaining, with professional stoicism, to take any
direct notice of " land lubbers."

Here we are, then, precariously making our way
along the rough yet slippery track, close to the edge
of the sea, throwing more attitudes than a posture-
master, in our efforts to maintain our footing on the
weed-draped stones. Now and then, despite our
efforts, down we go; or our unlucky foot slips in between two stones, and gets an awkward wrench; or,
recovering our equilibrium with a violent jerk, the
collecting basket flies out of our hand, and a jingle of
glass tells to the rueful ear that one of the jars is
smashed. Perhaps an old stager among sea-rocks,
who has bought experience with many knocks and
rubs and scratches, may venture to give a hint to the
novices among his readers, which may save them
many a slip. It is this: Never put your foot down
on a surface *that slopes away from you;* a slope
towards you is almost always safe. It is the angle of
inclination that makes all the difference; the extent
of surface is of little consequence, if it be large enough
just to receive the central plant of your foot-sole.

Here is a likely-looking stone; the great olive tufts
of saw-edged *Fucus* on whose back indicate that it
has lain in its present position for a season at least.
We put our fingers under its edge, and heave with a
will. 'Tis back-aching work, as you will find after a
couple of hours' earnest toil in a full noontide sun in
May; but over it comes, and displays a nice little
pool of clear shallow water beneath. A twinkling of
fins and tails; a splashing and a dashing of the
water; a hurrying of some dark slender object hither

and thither; kindle your enthusiasm. "We must have *that!* What was it? It certainly disappeared under this stone." "Well; lift it up and see." "Yes, there it scuttles away. Oh dear! 'tis gone under another stone." "Up with this too. Now then, be quick, or you'll certainly lose him." "There he goes again." By this time, however, we have had a pretty fair *glimpse*, at least, of our prey, that is to be. It is a slender serpent-like creature, mottled with shades of warm brown, and handsomely marked with regular square spots of deep black all down the long back. Is it a fish? or what? Yes, surely: it is the Spotted Gunnel, one of the Goby family of spinous-finned fishes, rejoicing in the aliases of Butter-fish, from its sliminess, and of Swordick, from its blade-like form; and variously called by Ichthyolo-gists, *Blennius gunnellus*, *Murænoides guttata*, and *Gunnellus vulgaris*,—which altogether make a by no means bad catalogue for a little fish to choose from, when he has occasion to sign his name.

As yet, however, we have only *seen* him, and that but momentarily: we want to possess him. This is a not very easy achievement, so nimble and evasive is he. He has learned also by this time that he is "wanted," and makes no secret of his preference for liberty. "I'd rather not!" is expressed in every slap of that little tail-fin, in every undulation of that snaky form. The channels between the stones communicate with one another, and the fish seems to have an intui-tive and instant perception in what direction the most

available way of escape lies; so that we almost despair
of securing him. At last, however, he has taken
refuge under the edge of a stone which bars up
egress. Cautiously we bring both hands to bear,
placing one on each side, so as to make a sort of
basin; then inserting the fingers under the shelter to
probe the retreat, out darts the little hider, but finds
himself environed in the hollow of our hands.
"Quick! the jar! or he'll escape yet!" No! he's
all right; safely dropped in; and we hold up the
glass vessel, half full of clear water previously
provided, and gaze in triumph at our little captive.

An elegant creature it seems as we now behold it,
gliding round and round the bottom of its crystal
prison; now turning lithely on itself, like a fold of
narrow ribbon, now swimming through the clear
water with the most elegant undulations, but sinking
to the bottom again the instant the undulatory effort
ceases (for it is strictly a ground fish), and manifesting
in every movement the perfection of agile grace and
elegance.

It is, as I said, a fish of a somewhat serpent-like
form, or rather ribbon-shaped, for its height much
exceeds its thickness. It is a ribbon set upon its
edge, with a fringe above and below, and each prettily
marked with regular chequered spots. But I must
jot down a more methodical description of him, that
you may recognise him again when you fall in with
him; and this you will probably do every time you

spend an hour stone-turning, at least on this south-western coast.

The length of this specimen is about six inches, and I have seen larger; the height or vertical diameter is about one-third of an inch, and the thickness, at its greatest, which is on a line midway between the back and belly, is about one-eighth of an inch. The body comes to a sharp edge above and below, and is fringed on both edges with a long narrow uniform fin. That on the under side, the *anal*, as it is technically termed, is found only on the hinder portion, commencing behind the middle; while the *dorsal* or back fin begins just behind the poll; both emerge into the *caudal* or tail-fin, with which the body terminates. The *pectorals* or breast-fins are very small, much like those of the eel, and the *ventrals*, or belly-fins, are so minute as to be scarcely noticeable, just a pair of tiny points. The head is small, with a smooth descending profile; the mouth very short and feeble; the eyes moderately full, round, bright, with golden irides.

In colour our little Gunnel is of a warm, yellow brown or olive, mottled with a delicate sort of curdling all over, with some dashes of reddish here and there; a band of deep brown descends perpendicularly down each cheek from the eye; the long dorsal fin is marked at regular intervals with square spots of black, bordered on each side with lines of pure white, adding greatly to its beauty. These descend a little on the fleshy part of the back. Similar spots adorn the anal, but being only of a slightly deeper shade of brown than the ground colour, they are less conspicuous. Altogether it is a pretty and attractive creature.

Meanwhile, however, we pursue our searching. Stone after stone we turn over, discovering many curious and interesting things — starry Tunicates, bright-hued Nudibranchs, errant Crabs, Stars, and Urchins, and Sea-cucumbers, and multitudes of other creatures well worthy of examination and preservation. Our capture of the Gunnel, however, makes us eager for *fishes*, and we look out for more. Nor in vain. Under a heavy block, which we succeed in overturning only after several slips, there lies a huge thick-set uncouth object, which makes such a splashing that we cannot distinguish his form at all. We have not much difficulty in securing him, for though he makes a considerable pother, and stirs up not a little cloud of mud from the bottom of the shallow pool, we easily manage to place our hollowed hands round him, and lift him bodily up. Very indignant looks he as we drop him into the glass jar; his eyes goggle unutterable things at us, and his muscular tail and broad caudal fin whisk from side to side with wrathful energy as he flaps to and fro, and round and round, making a tempest almost equal to that renowned one " in a teapot," and quite discomfiting the amazed Gunnel, his lithe and graceful co-tenant, who is fain to perform undulatory evolutions at the top of the jar, relinquishing the entire bottom to the boisterous intruder.

Its singular palmate and branching horns, remotely
resembling those of the fallow-deer, at once shew it
to be the Gattorugine, another fish of the same family.
Fleming, in his *British Animals* (p. 206), speaks of
it as British only on the authority of two specimens,
one recorded by Pennant, found near Anglesea, and
another mentioned by Montagu, as taken in a crab-
pot on the south-east coast of Devon. Yarrell, however, adduces other examples; and on the authority
of Mr Couch, speaks of it as common in Cornwall,
where it goes by the name of " Tompot." " Tompot,"
then, from henceforth, he shall answer to with us;
for hereabout it is one of the most familiarly abundant
of species, when searched-for as I am describing; and
" Gattorugine " is too recondite a barbarism for the
appellation of so everyday a subject as this. So, elaborate dandy as he is, " Tompot " shall be his name.
Be quiet now, Tompot! be still a moment, that we
may solace our eyes with your beauty!

Beauty, forsooth! A more hideous " varmint " you
will scarcely see, notwithstanding his gay attire and
his jewellery—for he wears both. What a face he
has! His head is thin and high, with a profile descending, I will not say perpendicularly, but, indeed,
not very far removed from it; a mouth opening with
a most repulsively impudent expression; two large,
prominent rolling eyes, of which the colour changes
notably while you look at them, now being partly
dark and partly white, abruptly divided by a transverse line, then the whole rapidly becoming suffused

with a dark, purple hue, so that you can scarcely discern any inequality. In the middle of the face, on each side, is a minute fringed process, and above each eye a much more remarkable one, standing up conspicuously, and resembling, as I have said already, a deer's horn. To be more particular, each appendage of this pair is a thin piece of cartilaginous skin, standing erect, and facing diagonally, both edges of which are cut into a number of sharp points, just as one might snip a piece of paper with fine scissors. They are white, irregularly blotched with red. The use of these fine ornaments I do not at all know; I have never seen the least motion in them, nor the slightest attempt to use them made by the numerous individuals that I have kept in my aquarium. The pectorals are unusually ample, and very fleshy; the ventrals are thick, slender, two-rayed, and white, and are set on just under the throat, where they form a remarkable feature as the fish grovels about. There is but a single dorsal, running along nearly the whole length, but it is distinguished into two portions by a difference in the height and structure of the fin-rays. All the fins are fleshy, and indicate muscular strength.

As to colour, our Tompot is a beau. He wears a suit of light grayish green, studded all over with minute black specks, and the body is crossed by about seven broad vertical bands of dark reddish brown, extending to the edge of the dorsal. The cheeks are marked in a somewhat mountebank fashion, with dashes and square clouds of rust-red, which hue also adorns the chin and the pectorals. Such, then, is the portrait of this, our second capture ; and now to work again.

Presently, a third appears. This time an imp, with a bluff, cod-like head, and as black as ink, turns up. He, too, shoots hither and thither, but with an energy peculiarly his own, as different from the splashing vehemence of the Tompot as from the agile activity of the Gunnel. We catch him without any difficulty, and in we pop him into the already occupied jar. Rather a risk we run, to be sure, in crowding three such creatures as fishes, and of this size too, into one bottle ; but *necessitas non legem ;* our other jars are already filled with other things, and for half an hour, till we can turn them out at home, perhaps they may manage. They do not seem very delicate in constitution, all of them readily accommodating themselves to captivity, and very amusing they are there.

But we have not yet recognised the stranger. I said he was black as ink. No, he is not. He is of a dusky gray, or ashy hue, very prettily marked all over, with a warm sepia-tint disposed in irregular wavy bands, which are dappled or clouded ; and which, upon the dorsal, pectoral, and caudal fins, are arranged very close to each other, and are particularly elegant. The front dorsal, for there are two, has its upper half of a delicate pale straw colour, into which the dark gray hue runs up in points between

the rays. But how strange is this! Was he not,
then, black before? How could we have mistaken
his colour? Nay; it was no mistake. See! he is
rapidly deepening in hue; a few minutes, and he is
black again, black all over, with the exception of the
yellow half of his dorsal, which remains unchanged,
and comes out all the finer for the bolder contrast:
not a trace of the mottled bands can now be discerned, except slightly on the fins; but the blackness, which has become intense, has now assumed a
superficial bloom or flush of fine indigo-blue, just
like an untouched plum, or black Hambro' grape.
This remarkable change of colour seems to depend
on mental emotions. Alarm or sudden terror will in
an instant cause the gray tint to be put on, while the
bloomy black indicates equanimity.

The form of the fish, particularly that of the head
and face, and the expression, are totally different
from those of the two others. He puts on a vacant
stare, quite idiotic in aspect, whereas both his fellows
display a sort of intelligence. His lips are thick, his
cheeks fat, and his eyes enormous and staring. His
length is about four inches.

This is the Black Goby. Now all these three
fishes are with us exceedingly common, notwithstanding that Yarrell pronounces the last two to be
very rare. At all seasons I can turn up any, or all
of them, at an hour's notice, on Babbicombe Beach,
at low water of a good spring-tide; and as they are
all three very readily preserved alive and in health,

and as they quickly become familiar in a tank, they are quite worthy of being sought after by amateur zoologists.

We carry ours home, and, committing them to a large aquarium, study their habits. At first they are shy; hiding under the rocky covers, and inexorably refusing to come into view. We must give them a few days to overcome their timidity; they will soon become familiar, indifferent, and even saucy. Then they are very amusing.

The Gunnel always retains, indeed, some traces of its timidity. It habitually chooses for its residence an overarching stone, or some broken shell of a large bivalve, under the shelter of which it can lurk, with its pretty yellow head just peeping out. If driven thence, you will see it gliding along in every corner of the tank, seemingly in all parts at the same instant; but in a couple of minutes you discern the little nose peeping from the old hiding-place just as before. To-morrow you will find it in just the same place, and next day again. Sometimes it swims with much elegance, but with effort, and in a leech-like manner, through the open water near the surface.

The Tompot bursts and blunders about in an uncouth, headlong way. If you approach the tank he will leave his retreat, and rush up to the glass, as if to challenge your intentions. Then, as if satisfied, he turns round with a swash of his tail, and bustles away into the cavern; whence he will presently half emerge, and lie close to the ground, with his great lack-lustre eyes rolling to and fro, or gazing stupidly upwards, just as I have sketched him.*

The Goby is the liveliest. He, too, is fitful and sudden in his movements. He will frequently rest his chin on a stone, and stare at you, as if his life depended on his making you out, alternately elevating and depressing his pretty dorsal. The ventral fins are very curiously united into a kind of funnel or cup; and this is evidently an organ of adhesion; a sort of sucker. Montagu suggests this use, but he had never witnessed it. His words are, " With respect to the union of the ventral fins, it would seem to be for the purpose of forming an instrument of adhesion; but in no instance have we observed that they adhered, either to rocks or to the bottom of the glass vessel in which they have been kept alive for several days."†

Probably this excellent naturalist failed to observe it from not being able to keep the fish long enough to wear off its first surprise and timidity, incident to its unwonted situation and circumstances. But I have seen the action repeatedly. The fish will frequently swim up to the glass side of the tank, and suddenly adhere in a perpendicular posture, remaining motionless for several minutes. During this time the curious formation of the ventrals may be well observed.

* See the Frontispiece. The Gunnel is represented as approaching in the centre; the Tompot projects his head from beneath a stone in the right corner; and the Black Goby is seen resting his chin on a block on the left.

† Montagu's MS., quoted in Yarrell.

The Goby feeds more readily than the others. Little atoms of meat dropped in soon attract his attention. He turns up his eye, and watches them as they slowly sink; then, when he judges the proximity sufficient, he shoots ahead, opens his cavernous mouth, and sucks in the morsel, and is in an instant ready for the next. You may teach him thus to come towards you when you approach. The Tompot will not so feed; but while the Goby is seizing the morsels, watches him, bustles up, and drives him away, but will not attempt to help himself.

Meadfoot and the Starfish.

T

MEADFOOT AND THE STARFISH.

DADDY HOLE.

Brilliantly and fervidly the sun blazed down on the laughing earth on the morning of what, in my early days, used to be designated the " glorious" 1st of June, as I wandered through fields, and over downs, and along the edges of turf-crowned cliffs; partly to see a remarkable chasm in the limestone on Daddy Hole Plain, near Torquay; partly to search for the yellow horn-poppy, a fine, and not common plant, which, I had been told, grew thereabout, and was now in blossom ; and partly to see what might be picked up in the way of marine natural history among the sea-washed rocks at the base of the precipice.

An hour's stroll, which produced nothing that I care to mention now, brought me to the plain, an elevated down, in which, near the seaward edge, I came suddenly on the yawning chasm. A great strip of the limestone margin has slipped, separating from the main body, and essaying its descent upon the far beach below. It has, however, been arrested in its course, and thus remains in its integrity,—saving some clefts and fissures,—but leaving between it and the mainland a great gulf, some thirty feet wide, and about sixty deep on the average. The sides descending perpendicularly, resemble rugged walls of the hard gray limestone, a resemblance heightened by the stratification, which is here quite horizontal, like courses of cyclopean masonry. Gray and black and orange-coloured lichens give their many tints to the harsh stone, and ferns and herbaceous plants throw out their luxuriant tufts of various shades of green from every crevice. Young ash-trees, bird-sown, springing from the débris at the bottom, have reared their heads of graceful pinnate foliage almost to the level of the walls, where the ivy, in its deep, dark, glossy verdure, drapes the edges, and hangs down in profuse festoons. In the intervals between, the margins are bright with the white rock-rose, the cheerful yellow lotus, and several kinds of stonecrop in fullest bloom. As I stand, gazing down, a blackbird, with shrill, clamorous scream, rushes up from the obscurity, and at the same moment a couple of butterflies—the pretty tawny, black-chequered sort which collectors call the " Wall"—rush out from some hanging herbage, and perform playful evolutions in the midst, their forms and colours seen in fine relief against the dark background of the chasm. White-rumped martins shoot to and fro through the narrow fissure, snapping up the humming gnats and minute beetles that are playing on the wing.

I walk along the margin, obtaining new and changing peeps of the depths, till I reach the southern extremity, where a rough, broken, zigzag sort of stair promises access. I essay it, pressing through the briery bushes, occasionally making some awfully long steps, and finding my heavy vasculum somewhat embarrassing, till, after some scrambling and slipping, I find myself at the bottom.

A feeling of solemn awe creeps over me as I stand between these rugged walls, in the bowels of the living rock. It is so still and silent, that the sound of my walking-stick, set down on the ringing blocks of stone, falls startlingly on the ear. I look up to the huge buttresses and angular groins of stone projecting into and narrowing the space, and admire the pellucid greenness of the out-springing ferns seen against the slender line of bright sky. Thridding my path among the fallen masses, I make my way to the other end, and manage to climb out by a devious fissure upon the sloping side of the sea-face, all strewn, and almost covered, with heaps of shattered fragments. These give place to broken but verdure-clad ground, where the pretty white rock-rose (*Helianthemum polifolium*), a somewhat rare plant in England, is growing in abundance, forming large cushions of neat hoary foliage, studded with hundreds of the delicate snowy blossoms, the crowd of stamens making a pretty centre of bright golden yellow to each. I go down on hands and knees, and labour to secure some roots for the rock-work of my garden. Nothing seems easier than to get them up out of this shaly loose stuff, but really it proves a trial of skill and patience. The surface fragments are soon scraped away; then you come to larger blocks imbedded in the clay; the long taproots run in between these masses, which have to be worked out; but the more one digs, the farther seem to penetrate these interminable roots, which have scarcely any rootlets till you come to their tips. However, I managed to secure a few fair specimens uninjured, selecting chiefly young plants, as being more likely to survive, and boxed them up in the vasculum.

While I am thus engaged, a cuckoo calls in startling proximity. It is evidently in a bushy hollow just below a knoll on which I am kneeling. I wish to get a peep at the bird; for, familiar as is the cuckoo's voice to our ears, a near glimpse of his person is by no means a common thing. I cautiously crawl over the knoll; the well-known call comes up again— "cuckoo!" singularly distinct. As I gaze, the bird sud-

denly spreads its blue wings just beneath my feet, and
shoots away, calling vociferously as it flies, the sound
becoming more and more faint and mellowed, till the
bird reaches a distant wooded hill. I never before
had an opportunity of seeing a cuckoo to the same
advantage; it is always an interesting bird to my
mind, and its elegant shape, and beautiful though
sober colouring, were well set off under the favouring
circumstances.

I sat down to rest on the brow of the steep seaward
slope. The sea lay in its vast expanse, magnificently
spread out below and before and around, recalling, as
it generally does, in its calmness, the sacred simile,
" like a molten looking-glass." Hope's Nose, on the
left, bordered the view. Berry Head, stretching far out
into the bright plain, like a vast breakwater, was on
the right. The ships and small craft, coasters, fishing smacks, yachts, and pleasure-skiffs, were speckling the glittering bay with their white sails, creeping
along under the light air which blew in soft gushes
off the land. A couple of war-steamers, black and
huge, with their grim rows of grinning teeth, were at
anchor off the Head—

> " Like leviathans afloat,
> Lay their bulwarks on the brine."

A shingle-beach, merging into some rugged, weed-
clad points of rock, stretched along at my feet, from
which the sibilant whisperings of the ripple came up
to the ear; and two or three rocky islets, rising in

black repose out of the glistening sheen, served as a foil to the mirrory brightness of the water.

Charming it was to sit and gaze on the lovely scene. The exertion of walking and collecting had given just enough of fatigue to the muscular system to make the *dolce far niente* a luxury. Under the shadow of a great angular block, I reclined, enjoying the beauty and exhilaration of the sunlight, while relieved from its oppression. Most brilliant was the flood of light with which every object was suffused in the unclouded blaze of that summer noon. How fine was the interchange of broad light and deep dark shadow, on those angular limestone cliffs! How glowing the coloured breadths of golden furze and purple-sheeted heath, expanded sea and vaulted sky! How soothing the warblings of the merry birds coming up from the earth on every side, and coming down from the beaming sky, where the dazzled eye vainly sought the soaring songster!

But up, and to work again. As yet I have seen no trace of the horn-poppy; though it was seen plentifully in blossom hereabout a day or two ago. I roam to and fro along the irregular slope towards Meadfoot, searching the different levels. Yonder is the fennel growing profusely. Ha! my informant told me that I should find the *Glaucium* near the fennel. Encouraged, I go down to the spot. More fennel! numerous tufts spring out of the rocky soil, the vegetation, in its smooth, green stems, and arching, feathery foliage, somewhat reminding me of the

noble bamboo-crowns that I have admired in the
tropics — of course, in miniature. Still no *Glaucium!* I peer about, and espy a scarped footpath,
partly broken away. I remember my friend mentioned this in his description of the locality. I jump
down, and in a moment see plenty of the desired
plant growing on the outer edge of the path, and
below it. Scores of specimens display their singularly rigid, deeply cut, and grayish-white foliage, their
large and rather showy yellow flowers, and their long
horn-like seed capsules. I find their deep roots even
harder to extract than those of the cistus, at least in
a condition which yields any hope of their surviving
the transplanting; but by selecting the youngest
specimens, I did succeed in boxing two or three, and
satisfactorily effected the transfer.

Thus two out of my three prescribed *desiderata*
were achieved; there remained the search for marine
animals in the rock-pools, and under the stones. I
was by this time not very far from the shore-level, and
presently stood on the beach. The tide was not in
the most favourable condition; and though the rocky
points that ran out into the sea hereabout were broken
enough, they did not evince much tendency to form
hollows and basins capable of retaining the sea-water
when the tide had receded from their level. But the
crevices between the blocks were well fringed with
red and purple sea-weeds, which waved elegantly as
the gently swelling sea lifted them up and down. The
Palmate Rhodymenia, sometimes called *dulse*, and

eaten by the poor people on the northern coasts both
of Scotland and Ireland, was abundant, forming fine
masses of broad, somewhat leathery, smooth leaves,
irregularly cut and split, of rich deep-red. The
Chondrus crispus, which, when dried and bleached,
is sold in chemists' shops, under the name of " Car-
rageen Moss," was also numerous, growing in stiff,
divaricating tufts of a deep purple-brown hue, each
narrow, strap-like leaf terminated by tips of the most
glowing, steel-blue lustre while under water ; a
beauty which, however, utterly vanishes as soon as
you remove it from its submersion. Along the edges
of the narrower fissures, between the stone blocks,
grew soft, plumose tufts of the Winged Delesseria, a
plant of very lovely colour, belonging to a genus,
every species of which is beautiful. It consists of
very narrow leaves, each composed of a mid-rib,
and a slender wing of membrane bordering it on
each side ; each is very much divided in the same
plane, and thus the whole constitutes a thick, bushy
tuft of a few inches in height. Viewed in the water,
its hue is a dark brownish-red ; but when looked at
with the light behind, as in a glass of sea-water, the
leaves are of a rich, light carmine. And far down in
these crevices, deep under water, I could discern the
large, sinuated broad leaves of the most splendid of
our native Algæ, the Sanguine Delesseria. This is a
far finer species than the other, though formed on the
same model ; a well-grown leaf is sometimes eight
inches in length, and nearly three inches in breadth,

consisting of a thick, firm mid-rib, with regular nervures spreading on each side, on which is stretched a
delicately membranous leaf of the richest transparent
crimson, finer than the finest cambric, most elegantly
frilled or puckered all along the edges. This very
fine species is not uncommon all along the coast hereabouts, but is never seen except at the lowest level of
the tide, where it grows often in considerable quantity,
large leafy tufts springing out of the basal angles of
the perpendicular masses of rock, or in persistent
tide-pools hollowed in the rock itself. It will not
bear exposure to the air with impunity, as many of
our sea-weeds will; for if left uncovered but a short
period, a quarter of an hour or even less, the delicate
rose-crimson membrane becomes defiled with large
blotches of a dull orange-colour, which shew that its
texture is irrecoverably injured, decomposition having
already set in. The disease inevitably spreads, and
in spite of all care the orange spots decay, and soon
slough away, leaving only the mid-rib, which gradually decays in like manner.

I tenderly lift the drooping fronds of the Rhodymenia, and the first thing I see is a huge Five-finger,
clinging to the rock with four of its thick fleshy arms,
while the fifth is raised into the free water, its tip recurved, and its scores of pellucid sucker-feet stretching
and slowly waving in all directions, as if exploring
for some new resting-place, or searching around for
some object of appetite. It is much too big to be
carried home as an aquarium-pet, but as there seems

little else to be obtained in this condition of the tide,
I determine to spend a half-hour in gaining a closer
acquaintance with it here. So I seize it, and dragging
it from its many-footed hold of the rock by main
violence, not without amputation of some of the
suckers, which will tear apart sooner than relinquish
their grasp, I bear it away in triumph, seeking a convenient theatre for the display of its powers and per-
formances.

Here is a little basin, chiselled, by frost and wave
combined, out of the rough rock; it is half full of
clear water, left by the retiring tide, but seems tenantless so far as animal life is concerned, though two
or three dwarfed and stunted *Algæ* are growing from
its edge and dipping their frond-tips into the water.
Into this I drop my captive, nothing loath; he sinks
like a stone to the bottom, alighting on his back; and
there he lies sprawling, with the suckers protruding
to their utmost, doubtless seeking to "realise" the
conditions of his new whereabouts. Soon one or two
of the arms begin to curve their extremities under, so
that the suckers can touch the rocky bottom; then
the curvature rapidly increases, and at length the
whole creature sluggishly turns over, and he is all
right again, *pied a terre.*

It is the common Starfish, Crossfish, or Five-
finger of the fishermen of various parts of our coast,
the *Uraster* or *Asteracanthion rubens* of zoologists,
the most familiar example we possess of the class
Echinodermata. It is, withal, one of our largest

species, for it sometimes reaches a diameter of more than a foot, with the rays nearly two inches wide. It has a somewhat coarse appearance, for though the colours are often gay,—red, orange, yellow, or purple above, and pale straw-yellow or cream-white beneath, —yet the surface is usually blotched in a scaly or leprous manner, and the semi-crustaceous texture of the skin, something between leather and shell, effectually precludes the idea of personal beauty. Still it is a very curious subject; and as I mark it gliding smoothly, and with a moderate rapidity, over the unevenness of the rocky bottom, and notice the mechanism by which its progression is effected, I see at once that I have before me one of the great types of animal locomotion; a series of contrivances, by which a given end, that of voluntary change of place, is accomplished, which are quite *sui generis;* admirable in their adaptation to the prescribed end, but totally unlike the arrangements by which the same object is attained in higher forms of life.

This is worth studying in detail. Here we have one of the multitudinous results of infinite Wisdom and almighty Power combined in creation. The problem is to endow with the faculty of voluntary locomotion a sentient creature which has no internal skeleton, and no limbs. It is solved in many ways in the invertebrate classes, and this is one example.

Each of the five thick and bluntly-pointed arms, or rays, of this star-like animal is seen to be indented on its under side by a rather wide and deep furrow, which extends from the hollow in the centre, where the mouth is seated, throughout its length, to the point. Along the floor of this groove we should see in the dead and dried animal four rows of minute perforations, running lengthwise. We cannot discern them directly during the living activity of the starfish, because the crowding sucker-feet conceal them. Each of these suckers is a tube of delicate membrane, a continuation of the common skin; and its interior accurately corresponds with one of these perforations in the skeleton. The tip of the tube is expanded into a broad circular flat disk, which retains its form, owing to its being strengthened by an internal horizontal plate of calcareous glass, which has a broad hole in its centre. The use of this you will presently discover.

If we were to dissect this animal, we should find, on the interior surface of the semi-crustaceous integument of the arm, a little globular bag of similar transparent membrane, on each aperture, which opens into the cavity of the globe, just as on the outer side it opens into the tube. Thus there is a free intercommunication between the globose sac on the inside and the sucker-tube on the outside, through the tiny perforation in the crust. The interior is filled with a clear fluid, scarcely differing in its nature from seawater. The globular sac within and the tube without are both composed of highly contractile tissue, under the control of the animal will.

We must bear in mind that we have been considering only a solitary example of these structures; but there are hundreds of them, all the exact counterparts of each other, wielded at the same moment by the Starfish. Thus the power exerted, feeble in each isolated case, becomes very great in its accumulation; and when the animal has brought a dozen or twenty of its suckers out, and having stretched them forward to the utmost towards a certain point, has firmly attached each, and begins to pull upon them, relaxing at the same moment those which have been attached behind, the whole body, heavy as it seems, is easily dragged up toward the point of resistance. The action is just analogous to that by which a ship's crew, having carried out a couple of anchors a-head, tighten the cables, and, by means of the windlass, warp up the vessel to the anchor's place. The Starfish's progression is a constant succession of such warpings, with the advantage of a large number of anchors and cables to work upon at the same time.

Thus I have got a good half-hour's entertainment and instruction out of my captive; and now, leaving him to comfort himself in his rock-pool till the quickly rising tide covers him once more, I wend my pleasant way homeward in the cool of the afternoon.

A DAY IN

The Woods of Jamaica.

A DAY IN
THE WOODS OF JAMAICA.

THE FORDS OF PARADISE RIVER.*

I⟮ is now nearly sixteen years since I came to the end
of that delightful sojourn in the forests of Jamaica,
which passed so swiftly, and which now seems like a
midsummer night's dream ; but often and often I say

* From a sketch of my own. The singular hillock-like objects
are bushes completely covered with the foliage of a species of *Con-
volvulus*, as if with a mat.

to myself, as I muse upon the past, and recall the beauteous scenes,—How I should delight to transport myself in a moment, as by an electric telegraph, to the summit of Bluefields Peak, or to the sombre glooms of Rotherwood, or to the sunny glades of the Kepp, and spend just one long day in re-exploring there! I shall never see them again in this body, for those sweet ties that advertisers strangely call "incumbrances" have clustered round me, and gray hairs are peeping out of my head and beard, and mere locomotion has not the charm that it once had; but I sometimes think that, when the Lord Jesus, bringing in the times of restitution of all things, shall clothe this sin-pressed globe with far more than pristine glory and loveliness, and I have put on the resurrection body, fashioned in His likeness, to whose incorruptible, immortal powers time and space will be as nothing, one of the myriad joys reserved for me may be the looking again upon those gorgeous scenes of beauty, which, even as I have already seen them, are so little marred by the sin of man, and retain so much of Paradise,—the mountain-woods of glorious Jamaica.

There may be a few of my readers, to whom it will not be disagreeable to accompany me in imagination, while I tax memory, and try to paint, in poor and feeble words, a few of the features and objects that would vividly strike a stranger's fancy, who had never before been out of Europe.

It is our first visit, then, to a tropical land. The ship that brought us hither has cast anchor in a bay

formed by the coral reef, on which the Wrasses and
Parrot-fishes, glittering in green and azure and scarlet,
are playing and nibbling, and the little butterfly-like
Chætodons, banded with black on golden-yellow, are
shooting to and fro. As we sit in the stern-sheets of
the gig in which we are pulled ashore, the brilliantly
transparent water reveals these and multitudes of other
things, enjoying themselves beneath us. Those noble
Cassides and Conch shells, which are the ornaments of
our sideboards at home, are seen gliding on the sand;
great purple urchins with enormously long spines, and
others with curiously broad flattened pegs of alternate
colours ; sponges, of forms imitative of umbrellas, and
cups, and goblets, and trees, from the size of a dining-
table downward ; gorgeous sea-slugs crawling on the
grass-wrack ; madrepores and corals clothed with their
bright-hued anemones ; snatches and glimpses of these
and hundreds of other strange creatures, nourished in
the steaming sea beneath a vertical sun, we catch, as
the boat shoots rapidly over the glassy wave ; and now
her keel gently grates upon the glistening beach of
coral sand, and we are ashore on Jamaica, the gem of
the Caribbean Sea.

The great south road here skirts the shore ; and at
this little wayside inn we procure horses and a negro
attendant for our mountain excursion ; for we cannot
walk far in this climate with enjoyment. While the
steeds are saddling we stretch ourselves on the sand
under the shadow of this sea-grape, whose round,
shield-like leaves are veined with richest crimson, and

whose long racemes of purple berries tempt the eye and the palate, and admire the tufts of white lilies, having the fragrance of carnations, whose bulbs delight to be washed by the wavelets of the sea.

We are mounted; and now let us rapidly get over the lowland slopes, to reach the loftier regions as early in the day as possible. The sun has not yet risen; and there is a dewy freshness in the air, as the dying land-wind of the night comes off in intermittent breathings, bringing the perfume of ten thousand flowers. Here, between cliffs of limestone, where creepers festoon the rock, and the noble trumpet-blossoms of the Portlandia, snowy-white, and each eight inches long, hang.down from the clustering foliage, out of every fissure,—we make our way up a steeply rising track. The cliffs on either hand soon begin to recede, and we emerge on a road between pastures of Guinea-grass, whose brightness never withers under the driest seasons. Orange-trees line the road, loaded with their golden fruit; and sops and custard-apples, and luscious naseberries and guavas, are scattered over the fields. Birds have awaked: the Petchary, earlier than the yard-cock, long ago piped from the fronds of the tall cocoa-palm; and yonder we see one continuing his simple song with unabated energy, opening ever and anon, as he shifts from twig to twig, the bright golden coronet upon his head. Ha! he is not doing that for nothing. It is the expression of excitement. He has ceased to sing; watch him! A large beetle is crawling near, which is in the act of spreading its wing-sheaths for flight. Off it sails on drony wing! the Petchary instantly makes sail too; catches the heavy prey, and bearing it in triumph to his watch-post, beats it to pieces with his strong hooked beak, and swallows it.

Sweetly from the tangled woods of yonder hill issue the mellow notes, soft and broken, of the Merle; you would think it your own familiar blackbird, by the note, and would scarcely be undeceived by a sight of the bird itself; but it is a species peculiar to us. What we here call Blackbirds are larger birds, allied to the cuckoo; impudent, clamorous, sociable creatures, with a noisy, intrusive cry, like "Going away! going away! going away!" as they sail along on short, heavy wing, and long, balancing tail, close to the ground. There! we hear of a flock of them now; and yonder they are in the cattle-pasture, blackening the ground. They are cutting the droppings through and through, searching for maggots and worms; and for this purpose they are provided with a very deep, knife-like ridge on their beaks, which serves them as a ploughshare. See, too! on the backs of the patient kine, and clustering around their feet, are other sable attendants; sable they look from hence, but if we were close, we should find them adorned with the richest steely purple and blue-green reflections. With what business-like earnestness each searches among the hair of the cow he has selected to patronize, digging for bots and ticks; or walks round and round, with the ivory-white eye turned up, scrutinizing the grazing beast beneath, and now and then springing upward to seize the insect prey. Away goes one, the boat-shaped tail folded on itself, with a sharp metallic cry, which reminds you of the smitings of a smith upon his anvil. From this sound, we call the familiar bird the Tinkling.

As we proceed we hear the low sweet cooings of hundreds of Doves of various species coming from the woods. These sounds are eminently characteristic of the early day, in these wooded slopes. The loud and vehement call of the White-winged Turtle-dove, "Two *bits** for two!" is pertinaciously uttered; or now and again exchanged for its stammering cry of eight notes, of which the last is protracted with a moaning fall. The Pea-dove shews its plump form of purplish-fawn colour, and its large melting gazelle-eye, on the road before us, dusting itself almost under our horse's feet, or sits in the shadow of the groves, and coos, " Sary coat true-blue!"   And Ground-doves, no larger than sparrows, congregate in small flocks on the pasture-lands, searching for seeds of grass and weeds, and shout "Meho! meho!" or a loud and hollow "Whoop!"

The birds have their proper regions. We are attaining a considerable elevation, and are passing, by a narrow path, through a dense copse of small trees; bastard-cedar and logwood, with fiddle-wood and mahogany, much interlaced with briers and twiners.

* A "bit" is the colonial appellation of three-halfpence sterling. The renderings of the birds' voices are in *Negro-English.*

Here we see the beautiful and gentle Whitebelly Dove
walking to and fro in the greenwood shade, picking
up fruit-seeds; its light-coloured plumage rendering
it conspicuous; while from all sides the mournful
sobbing notes of this species resound.

The negroes delight to ascribe imaginary words to
the voices of birds, and indeed for the cooings of
many of the pigeons, this requires no great stretch
of imagination. The beautiful Whitebelly complains
all day, in the sunshine as well as the storm,—" Rain,
come, wet me through!" each syllable uttered with a
sobbing separateness, and the last prolonged with such
a melancholy fall, as if the poor bird were in the extremity of suffering. But it is the note of health, of
joy, of love; the utterance of exuberant animal hap-
piness; a portion of that universal song wherewith
" everything that hath breath may praise the Lord."
The plumage, as usual in this family, is very soft and
smooth; the expression of the countenance most en-
gagingly meek and gentle. And it is a gentle bird.
I have taken one into my hand, when just caught in
a springe, full grown and in its native wildness; and
it has nestled comfortably down, and permitted its
pretty head and neck to be stroked, without an effort
to escape, without a flutter of its wings.

A short turn of the path brings us out of the wood
upon an open plateau, whence the eye commands a
noble view of the coast for many a league, and of the
silvery Caribbean Sea stretching away to a far distant
horizon. The sun is just rising out of a bank of reful-

gent clouds, and shedding a transparent glory, a sort of veil of golden gauze, on every object within reach of his ray. Hundreds of winged songsters tune their throats to welcome his beam; the glittering lizards peep forth, and bask in the crevices of the rocks; and bright-winged butterflies leap into activity, and dance over the blushing flowers.

How gorgeous is sunrise, when viewed from a commanding elevation in the tropics! And perhaps under no circumstances is it more beautiful than when seen from the verdant mountains of an island, where the emergence of the orb from the glittering ocean can be commanded, while the immediate surroundings are those of the forest and the peak, where the effects of the slanting sunlight on the varied foreground, the green and brown trunks of the columnar trees, the broad masses of foliage, and the gloomy recesses between; the many-coloured rocks breaking out, with their festoons of verdure; the gay insects and birds and flowers,—give fine contrasts and harmonies; and where the crowns of the loftier hills, the pointed peaks, lighting up with sudden purple and gold, give an imperial magnificence to the prospect.

Look at this ancient Silk-cotton tree! what a fine object is it, illumined in the morning sun! The enormous perpendicular spurs stand out like radiating walls from the huge trunk, looking almost as white as marble in the bright light, and throwing the recesses into dark shadow. Trace up the vast pillar-like trunk! the eye wanders up a hundred feet before

it detects a branch to break the uniformity of its
column; there the huge boughs spread horizontally,
each one a vast tree for bulk and extent. What an
aspect of strength in those contorted and gnarled
limbs! How far away they carry the umbrageous
canopy of foliage! And see, too, what a microcosm
is such a tree as this! The hoary trunk is studded
at intervals with tufts of parasitic plants of the pine-
apple tribe; these are called Wild-pines; they do not
bear eatable fruit, but their blossoms are often of
great splendour. There is one now in flower: from
a tuft of rigid arching leaves, which form sheathing
cylinders at the base, springs a fine spike of closely-
set flowers, of the richest purple and crimson dyes.
Another kind has the sheathing leaves more compactly overlapping in a sort of herring-bone or zigzag fashion, whence projects a longer, looser, and
more branched raceme of scarlet and yellow blossoms.
There are many not now in flower, for they vary in
their season of blooming, but the leaves shew that
they differ in species, though they possess a general
family resemblance. One sort, common enough, is
not at all ornamental. The negroes call it " old man's
beard;" the stems are very long, and as slender as
wire, which form great ragged pendulous tufts, of a
dull hoary gray hue.

And there, in the forks of the huge limbs, grow
enormous matted masses of various vegetation, too
remote from our eyes to be identified in detail; but
we discern bunches of orchideous blooms hanging in

the air; and feathery ferns arching out their elegant
tracery; and creepers running along the boughs, and
what look like tussocks of wiry grass at intervals, but
which are small tiny-flowered orchids, and long, long
ends of green twine hanging many yards in length,
now looped up in a loose bight, now swinging in the
wind in mid-air, now almost touching the earth, and
dividing at their extremities into three or four smaller
threads.

Here we leave our steeds, and penetrate these lofty
woods. How solemnly still is the air! A subdued
green light, like that of an ancient cathedral, is diffused, to which our eyes are scarcely yet accustomed.
The huge old trunks look like the pillars of the Gothic
fane, and from far up in the groined roof come
dancing beams of bright light, green and yellow and
crimson, where the sun's ray falls on a single leaf or
flower, that remind one of the stained glass of lofty
windows. The butterflies are gaily flitting about the
margin of the forest; for they are children of the
sun; the flowers, too, are there; the shrubs are tall
and close-leafed, and covered with varied blossom;
but neither insects nor flowers venture far within the
gloom of these primeval woods, save here and there,
where openings in the leafy roof admit the bright
sunbeams, and make a little parterre on the floor.
But delicately cut Lycopodiums of the tenderest green
creep over the ground, like a soft Turkey carpet,
thrown over everything; gnarled roots, out-crops of
rugged stone, fallen trunks and branches, the trophies

of the hurricane,—all are overspread with this verdant cloth, which softens all the angles, permitting the general outlines to appear, but concealing the roughnesses.

Ferns, too, delight in this softened light, and they are here by myriads. Here, at the base of a giant fig-tree, is a noble crown of the Golden Phlebodium, with its elegantly pinnate fronds arching widely over our heads; beautiful are its thick twisted rhizomes, covered with golden hair that glistens with satiny lustre, and the delicately fine rootlets, that cling to the gray roots of the tree, crossing and re-crossing, and winding over them, like a tangled web of brown thread. In the crevices of the rocks are tufts of lovely light-green Maidenhair, some more minutely delicate than our own Devonshire species, others with noble trapezoidal pinnæ of large size, which gracefully diminish in regular graduation to the tip of the pointed frond.

And here are ferns which are altogether strange to a European eye—climbing ferns, whose slender scaly stems, something like the body of a snake, run up the lofty trees, clinging to the bark, fringed throughout their irregular windings with small rigid fronds, like the oval leaves of a cranberry or myrtle.

Stranger still than these are the noble Tree-ferns. Here we approach a part of the humid forest, where these forms are characteristic. Slender stems, as straight as an arrow, thirty feet high, but no thicker than a man's thigh, covered with the diamond-shaped scars of old fallen leaves, set in regular diagonal arrangement, bear at their summit a gracefully swelling crown of leaf-bases, which expands into a wide canopy of minutely divided foliage, each frond a facsimile of one of our familiar lady-ferns, or shield-ferns, or brakes, immensely developed. Other kinds, of somewhat inferior altitude, but of equal expanse, are beset with formidable spines on the upper part of the stem, while the lower part is wholly encased in a coating of wiry rootlets, black, interlaced, and ever wet with the condensed moisture of these humid woods.

Yes, everything here is saturated with moisture. The air has a warm steamy feel, like that of a wash-house; you cannot, indeed, *see* the clouds of vapour, because there is no surrounding cold air to condense it, but you *feel* it, and breathe it—soft, clammy, heavy. The mosses and lycopodiums, when trodden on, are like soaked sponges; streamlets trickle down the rough trunks, and the great hollow leaves of the arums and caladiums hold water in their hollows, not only in silvery globules, but in cupfuls, clear and sparkling. The sheathing bases of the wild pines that grow everywhere upon these tree-stems are reservoirs of water, in whose genial depths the great painted Tree-toads lie all day, bathing their naked limbs, and from whence they utter those startling shrieks, or moaning, gurgling objurgations, which by night terrify the superstitious vulgar with visions of the dreaded Duppy.

We cannot proceed far into the heart of the woods
without a special provision of hatchets or machettes to
hew for ourselves a passage; for the twisted lianes,
like cords and cables, stretch from tree to tree, inter-
lacing and forming treacherous loops and nooses;
and many of these are so terribly spinous that they
cannot be touched with impunity. Fallen trees lie
in every direction, presenting thickets of branches;
or the trunks are so decayed, that when you have
climbed on one to get over, it gives way beneath
your feet, and lets you suddenly down into a grave of
saturated rotten wood. The Duck-ants, or Termites,
have built their great earthy nests, like barrels, here
and there, and you are in perpetual peril of treading
on their galleries, and bringing an army of ferocious
biting insects upon you; or you sit down to wipe
your perspiring brows, and in an instant are covered
with a host of great Coromantee Ants, more furious
than lions, the nip of whose immense jaws is enough
to throw a man into a fever.

We are not yet at the top of this mountain-range.
Let us return to our horses, and go higher yet. Cool,
delightfully cool, as contrasted with the parched
plains, are these elevated regions; and very pleasant
is a ride up this narrow bridle-path, turning and
winding to ease the ascent. Now we break out on a
new view of the silvery ocean sparkling under a vertical sun; now we obtain a magnificent glance over
the mountain-forest, and over successive ranges of
hills, range beyond range, far into the interior. Now,

again, the prospect is bounded to a few yards : on one
side an almost perpendicular limestone cliff, covered
with gorgeous-flowered twiners, and gay with butterflies and gilded beetles ; with prickly aloes, and dwarf
fan-palms, growing in the crevices ; and on the lower
side a wilderness of bush, with here and there the
broad flag-like leaves of a banana, or the glorious
pyramid of pink blossom of a mountain-pride breaking the uniformity, or the slender stem of a cabbage-
palm lifting its shining crown loftily above the mass.
Now the forest boughs are meeting over our heads,
shutting out the sky, and making a grateful greenwood shade ; and now we are passing along one of
those pleasant bamboo walks that are so characteristic
of the steep mountain roads in this island. These
have been planted by man, for the preservation of the
roads, which are scarped out of the rocky face of the
hill. The gradual disintegration of the exterior edge
of the road from external causes, such as the weather
and the wear of travelling, would soon destroy its
level, and necessitate the cutting of it afresh. To
prevent this, it is found sufficient to lay down lengths
of green bamboo just below the edge of the road,
along the mountain side, and cover them with earth.
These germinate at every joint ; roots strike into the
earth, binding it firmly ; and a rampart of young
shoots springs up, which, increasing every year in
number and size, effectually prevent the crumbling
away of the edge, and by throwing their feathery
arches over the road, form beautiful green avenues,

under whose grateful shadow the traveller may journey for miles, and scarcely feel the toil of the steep ascent.

Here, where the calcareous rock recedes from the perpendicular, and forms a steep slope, allowing the growth of trees, though the ground is covered with irregular blocks of broken stone, elegant species of shelled mollusca, snails, and similar creatures, abound. A limestone region is essential to the abundance of these animals, because it enters so largely into the composition of their shells. As we ride by, we see the beautiful shells, many of them of porcellaneous polish, and exquisitely painted in bands and stripes of colour, or most delicately sculptured,—clinging to the leaves of the trees; and if we were to turn over the loose stones we should find them in wonderful variety and number, sheltered from the heat of the sun in those cool and moist retreats; for, as with our own homely kinds, night is the appointed season of their activity.

Yonder floats by a flock of Parrots with a most abominable combination of harsh screams. It is the Yellow-bill, intent on a new feeding-ground. Like an immense Indian shawl spread in the air, the compact flock speeds by, all on the same level, but undulating; as each bird presents a plumage of golden green, with azure wings, and scarlet tail-webs, the sunbeams playing over the array of colours has a charming effect. There the bright cloud settles on a cordia-tree, whose profuse scarlet berries give a ruddy hue to it even at this distance; but which are destined to a speedy dissolution in those greedy gizzards. Silent as death are the squalling birds, now that they are on the food-tree, and if we had not seen them alight, we should not suspect their existence there. Our steeds are wearied with the six hours' ascent, and here we attain our utmost elevation. Leaving them to regale themselves on the juicy bread-nut leaves which faithful Sambo will pluck for them, and leaving him to enjoy the siesta which he will then gladly take

——" patulæ recubans sub tegmine fagi,"

or whatever he may consider as the proper equivalent of the classic *fagus*, we will make our way into a sweet glade, so solemn, so still, so lonely, so cool, so bowery, so delightful to every sense, that you will confess it is worth the half-day's ride to have visited it.

It is a narrow ferny lane, shut in by blossoming bushes, with the forest-trees growing a few yards back, and screening us with their towering foliage from all but a gleam or two of quivering sunlight. The noble, reed-like leaves of the Indian-shot throw up their scarlet spikes, and bunches of fantastic orchids are drooping from almost every tree. Here is the Fragrant Epidendrum, filling the atmosphere with the perfume of its curious white blossoms, which are, too, very pretty, the lip shaped like a deep spoon, and its waxy whiteness picked out within in crimson lines. See, too, that compact mass of rich violet bloom, that projects from a tuft of leathery leaves low down on the trunk of that small lancewood : it is the Ionopsis, a lovely orchid that affects these dense woods. There also the Lycaste that bears the name of Lady Barrington, whose waxen flowers of creamy white stand out from the plexus of winding roots below the bulb, is seen abundantly on the low trees, hardly aspiring to rise above the ground : while the beauteous crimson Broughtonia, one of the most charming of all our orchids, frequently seats itself among the boughs of some lofty fig or Santa Maria, some eighty or a hundred feet above the spectator.

I called this lane lonely. Nay ; for it is populous with gay life. One of the very loveliest of birds, not of Jamaica only, but of the whole world, makes this secluded spot his most chosen resort. Look along the avenue ! Why, within a score yards, there are a dozen humming-birds in sight at the same moment ; and what humming-birds ! They are all of the same species—the very gem of our ornithology ; Polytmus, the long-tailed. Brighter-coloured kinds you may find in Bolivia or Peru ; but for elegance of form, combined with tasteful beauty, I think our little friends here can seldom be excelled. As they flit to and fro, visit the flowers, disappear within the shadowy woods, shoot again into the sunlight, hang on invisible wings over a blossom, probe it for the nectar, cling to its corolla with the tiny purple feet, dart out after a gnat, dash at a rival in the air, whirl round and round in playful combat, return to the flowers, suck, and suck again, they give us ample opportunity to approach them, and mark

their beauty, their vivacity, and their minuteness.
The velvet-black hood, the golden back, the lengthened pair of sable streamers behind, the long coral
beak, and, above all, the gorget of the most lustrous emerald radiance, changing to black by the
slightest alteration, then flashing back the gemmeous
light again—how lovely are these! And the beauteous little creature is so fearlessly familiar, confident,
perhaps, in its locomotive powers, that we may come
close up to it, as it sucks, without alarming it.

Did you mark that long solemn note? There!
another! and another!—each just two tones below
its predecessor; each sustained like the notes of a
psalm, clear and sweet as the sounds of a flute. There
sings one of the most eminent of our woodland vocal-
ists, the Solitaire, rarely heard except in the loneliness
of these high elevations. The Spanish colonists used to
say it chants the *Miserere*. So sweet, so solemn, so unearthly are the notes, recurring at measured intervals,
and uttered by an almost invariably invisible per-
former, that the mind is remarkably impressed; and
it would require little tendency to superstition to induce the belief, in a stranger who heard it for the first
time in these majestic solitudes, that he had heard the
voice of an angel.

But evening is drawing on apace: the sun is fast
declining, and we must leave these charming scenes.
Let us begin to descend. Evening merges quickly
into night in these latitudes.

The Blue and the Bald-pate Doves are flying over

our heads in little parties, each kind seeking the fre-
quented roosting-tree; the Jabbering Crow flits along
with its strange guttural talk; the great Potoo hoots
from yonder stump; the White Owl shrieks in the sky.
Now the loud harsh screams of the Aramus pierce the
wood, coming down from the stony hill-side; the
cracked voice of the Gecko proceeds from the hollow
tree; and the shrill metallic note of some Tree-frog,
singularly sharp and penetrating, rises from every part
of the woods below us.

The Night-blowing Cactus is opening its large and
beautiful disk of petals like a sun, and its fragrance is
almost overpowering. The perfume of a thousand
other flowers is now brought out by the falling dew;
and large dusky moths are hurrying to and fro to
enjoy their nectar. And now, queen of night, the
moon arises; and scores of wakeful Mocking-birds
salute her beam, and begin their rich and varied notes,
which are to fill the night with music. Fireflies are
shooting through the glooms, making lines of ruddy
or green light, or glowing like torches as they sit upon
the dewy leaves.

We are again on the shore. Beauteous Jamaica,
good-night!

Ferns.

FERNS.

INTERIOR OF A FERNERY. *

IT is no wonder that the cultivation of Ferns is increasing in popular esteem. The wide, and ever

wider, diffusion of the principles of correct taste is
training the eye of the multitude to discern other
charms than those of gorgeous colour; and to seek, in
flowing lines and graceful curves, in minutely fretted

the artist and engraver have bestowed much care upon the subject,
and the drawing on the block was corrected by myself. In the
left-hand corner of the foreground is *Drynaria quercifolia*, not
having yet formed the oak-shaped barren fronds, which impart
such a charm to this fine species. In the centre of the foreground,
on the wall of the tan-pit, is *Darea vivipara*, with drooping, minutely-divided fronds; and behind this, still on the pit-wall, is that
very remarkable fern, all bristling with black hair, *Dictyoglossum
crinitum*. To the left of this, on the end-wall of the pit, is *Campyloneurum phyllitidis*, with erect strap-like fronds, much like
those of our Hart's-tongue. Over the head of this are seen the
fronds of *Pteris quadriaurita*, var. *argyrea*, with a band of snowy
white along the centre of each pinna, seated at the end of tall
slender footstalks, which grow from a pot standing on an inverted
pot, almost concealed by the *Drynaria* first-named.

Occupying the right front corner of the picture (but beyond the
pit in actual distance) is a large crown of fronds of the majestic
Drynaria morbillosa; beyond this, close to the door-sill, is *Tham-
nopteris nidus-avis*, the Bird's-nest fern, a very noble form; and
above the former, on a block hanging against the side of the door,
is the Elk-horn, *Platycerium alcicorne*.

Beyond the tan-pit two sloping buttresses project in succession
from the back-wall towards the middle of the house. They are
built of brickwork, hollow, and pigeon-holed; and are so thickly
stocked with ferns and clad with Lycopods (to the number of about
a hundred and fifty species), that only their outline can be detected
in the engraving. There are, however, some very fine tufts of
Gymnogramma chrysophyllum, which may be seen between the
Pteris and the *Dictyoglossum;* and some fronds of a large creeping
Hypolepis, (? *tenuifolia*), arch from the summit of the nearest buttress, near the glass roof.

The pendent baskets in front, contain Orchids (*Stanhopeas*); but
near the door are seen various small ferns in suspended baskets,
and among them a *Nephrolepis* in a cocoa-nut shell.

outlines, in slenderness which is not weakness, in
verdure ever soft and fresh and tender, an exquisite
delight which is perhaps more refined than that which
is found in flowers, however rich, however lovely. I
remember, in my earlier horticultural days, the remark
of a lady, honoured in memory now, but quite of the
old school, expressing wonder that I should like to
have "fern" in the garden. To her eye the *filix mas*
was as the *filix fœmina;* the *Polystichum* as the *Lastrea;* it was "fern;" not "*a* fern," but "fern" in the
abstract, identified with the acres of brake she had
been used to see in her youth, and esteemed as vile as
the *vilis alga* of the poet. But her youthful ideas
were imbibed nearly a century ago ; and doubtless, if
she had lived to these days, she would have learned
to inspect, with discriminating delight, the varying
filagree work of the many fronds that arch over her
friend's cherished ferneries, and to watch their development with an interest little inferior to his own.

It has been well observed, that Ferns are always in
bloom. Winter and summer are alike to most of the
stove and greenhouse species; and many of our native
kinds retain their leaves through the winter, with
their lustre heightened by its fogs, and scarcely dimmed by its frosts. To a conservatory or hothouse,
Ferns lend a peculiar charm ; the exquisite lightness
and grace of their forms combine with their evergreen
verdure to relieve, and so to augment, the effect of even
the aristocratic Orchids. A constant interest attaches
to them : new fronds are protruding their curled

heads from the damp soil ; the adolescents are expanding their hundred arms; the mature are displaying
their curious and beautiful fructification, or forming
young offspring-plants to dangle in the air at their
tips,—so that the pleased culturist has ever something
to explore, something to admire.

To see our native species to advantage, let a stranger of refined taste roam amidst the tall-hedged lanes
and "ferny combes" of sweet Devon, in whose mild
and moist climate somewhat of the balmy breath of
the tropics is inhaled, brought to her shores by the
impinging waves of the mighty Gulf-stream. Here
the Ferns attain a magnificence of dimensions and a
permanence of freshness seen in scarcely any other
district of the land. I have in my eye at this moment
a lane, one side of which is formed by a bank very
nearly perpendicular, and about fifteen feet high ; the
whole face of this steep is densely clothed with the
Hart's-tongue, whose glossy green fronds, two feet in
length and four inches in breadth, spring out in the
most beautiful arches from the top to the bottom,
many of them strangely crisped and curled, and many
displaying that tendency to multiplied fission which
forms so interesting a feature in this species, wherever
it is found in luxuriance.   Not far off is an old
wall, the upper half of which is one unbroken sheet of
Trichomanes Spleenwort, intermingled with Ceterach
and Wallrue, the tufts springing out of the old decayed
mortar so close together as to make a continuous
shaggy surface of the elegantly fringed fronds.   This

same lane presently leads into an open wood, where, beneath the great timber-trees, sparsely scattered, enormous crowns of the Male-fern and the Dilated Buckler-fern grow up on all sides, forming vast basket-shaped hollows of seven or eight feet in diameter, of which the individual fronds attain a length of five feet by actual admeasurement.

Or, let him visit sweet Killarney,—that lake of renown, which is perhaps the most lovely little bit of scenery in the whole of the three kingdoms ; and, making his way along the sinuous channel beneath the towering Eagle's Nest, wakening the lingering echoes as he goes, emerge into the wild Lake of Muckreep, and land where the flashing cataract pours down the waters of the Devil's Punchbowl from a height of seventeen hundred feet. Here, growing on the Turk rocks, ever wet with the spray of the falls, he will see in abundance the filmy pellucid fronds of the Irish Bristle-fern ; the broad, triangular laminæ luxuriantly depending from their thousand tangled rhizomes, as if fastened with iron wire to the slippery shelves and ledges.

Or, let him follow in the steps of Mr Foot, and ex-plore the wild glens and vertical clefts in the lime-stone rocks of West Clare, where the Frail Bladder-fern grows to an unusual size, and contrasts, in its peculiar green hue and delicate texture, with the bright colours of *Gentiana verna* and *Geranium san-guineum*, and mingles with whole sheets of the rare *Dryas octopetala*. Here the Marine Spleenwort grows out of the vertical fissures, and reaches the almost incredible length of three feet in the frond, in which condition it is actually indistinguishable from that West Indian form, *Asplenium lætum*, which has been considered a distinct species. The scaly Ceterach, too, makes great tufts, with the curious sinuous leaves fifteen and eighteen inches long. But above all, it is the region of the most charming of ferns, the lovely Maidenhair. "I cannot describe my delight," says the enthusiastic explorer, "when my friend brought me to this spot [limestone cliffs near Ballynalackan.] The inland cliffs are formed of horizontal beds of limestone; and on the vertical face of these cliffs, in the clefts or interstices between the beds, this most exquisite of all the ferns grows in its glory. In fact, for a distance of fully half a mile, if not more, the stratification of the rock is distinctly marked by the peculiar green hue of *Adiantum capillus Veneris*. Between this and the sea, almost every vertical fissure in the flat bed of rock over which we walked was filled with this fern; and on the sea-side of the road it is associated with the gigantic *Asplenium marinum* above described. All the Wardian cases in Great Britain might be well supplied with *Adiantum capillus Veneris* from Ballynalackan; and what was taken would hardly be missed." *

The localities of this augmented luxuriance in our native Ferns,—the extreme south-west corners of the two islands,—are indications of a characteristic pro-

* *Proc. Dub. Nat. Hist. Soc.* December 1859.

perty in the elegant tribe, their predilection for warmth and moisture. In cultivation it is found that most of our native species respond to a little cosseting in these particulars; and in the germination of the spores and the forwarding of the early development, perhaps all are sensibly benefited by such means. Islands in the tropical regions present these conditions in highest combination; and it is in such places that ferns chiefly abound. Clare and Kerry in Ireland, and Devon and Cornwall in England, are the regions which make the nearest approach to the conditions of a tropical island, stretching out as they do into the ocean towards the south-west.

Statistical calculations have been made by botanists which curiously illustrate this tendency. In Egypt there is but 1 Fern to every 970 flowering plants; in Greece, 1 to 227; in Portugal, 1 to 116; in France, 1 to 63; in England, 1 to 35; in Jamaica, 1 to 9; in the Sandwich Isles, 1 to 4; and in Raoul, a little island in the South Pacific, 1 to 1,—the Ferns forming actually half the vegetation; so that every second plant is a Fern.*

* Dr Hooker, "On the Botany of Raoul Island," in the *Proc. Linn. Soc.*, vol. i., p. 125. Some observations of interest on the geographical distribution of the ferns in this island I quote:—"The absence of any ferns (with a single exception) but such as are natives of New Zealand, is a very striking fact, both because the list is a large one for so small an island, (twenty-two species,) and because, if their presence is to be accounted for wholly by trans-oceanic transport of the species, the question occurs, Why has there been no addition of some of the many Fiji or New Caledonia island-ferns, that are common tropical Pacific species, the Fiji Islands being

The lovely island of Jamaica is very rich in Ferns;
and perhaps there is no spot in the world where this
charming tribe could be studied with more advantage.
Situate within the tropics, an island, but of considerable area, and divided by a backbone of lofty mountains, it possesses a great diversity of climate, and
a magnificent vegetation. I have already said that
every ninth plant is a Fern; but even this statement
scarcely gives an adequate idea of the extraordinary
profusion of the tribe in the richer and more favourable districts, such as the gorges, glens, and humid
thickets of the lower ranges of mountains. Perhaps
it is not too high an estimate to suppose, that of the
species of Ferns at present in cultivation in our English
hothouses and warm-ferneries, more than one-fourth
have been drawn from this tropical island.

only 700 miles north of the Kermadecs, and New Caledonia 750?
The only fern which is not a native of New Zealand is the Norfolk
Island *Asplenium difforme*.

"Still more remarkable is the total absence in the collection of
any of the plants peculiar to Norfolk Island; for Raoul Island is in
the same latitude as Norfolk Island, is exactly the same distance
from New Zealand, and the winds and currents set from Norfolk to
Raoul Island; in short, though the northern extreme of New Zealand, Norfolk Island, and Raoul Island, form an equilateral triangle,
with the exception of *Asplenium difforme*, there is not a single fern
of Norfolk Island found in Raoul Island, that is not also found in
New Zealand; whilst of the twenty flowering plants of Raoul Island,
no less than six are absolutely peculiar to New Zealand and Raoul
Island, and, with the exception of the tropical, widely-diffused
Pacific species, there are no phænogamic plants or ferns confined to
Norfolk Island and Raoul Island. It is further remarkable, that of
the Raoul Island ferns, *Cyathea medullaris* and *Pteris falcata*
have not been found in Norfolk Island."

Let us transport ourselves once more, in imagination, to the south side of Jamaica, and, sauntering along a devious way through the cane-pieces, and park-like pens, and pimento groves of the lowland slopes, gradually ascend the mountain-range that borders the coast, and track the clefts and gullies of the limestone formation filled with luxuriant and gorgeous vegetation, till we reach the silent, sombre forest that clothes the shaggy peaks. We land; and, before we have traversed a furlong, we are struck with a gigantic Fern crowding the morass that steams just within the belt of mangroves, whose bow-like feet are bathed in the sea. It is *Acrostichum aureum*: the fronds to the length of eight or ten feet rise in noble crowns, and arch outward, covering a vast area, their form elegantly pinnate, like the leaves of the ash-tree, of a bright green hue, smooth and glossy; and the fertile ones entirely covered beneath with the golden-brown sori, lost in one common confluence. The base forms a short, thick stem, rising out of the stagnant water, and it may therefore be considered a tree-fern of low stature. Further on, in drier spots, we see several species of *Elaphoglossum*, a name given in allusion to our own " hart's-tongue," to which they bear some resemblance in their simple lance-shaped fronds, such as *E. conforme* and *E. latifolium*, dwarf ferns with thick glossy crowded leaves, golden beneath: some, however, as *E. squamosum*, are narrow and strap-shaped, and have a singular appearance from the edges of their leaves and

Y

their long foot-stalks being thickly fringed with bristly brown scales. Then occurs a noble species, *Gymnop-teris nicotianæfolia*, of which the leaves, composed of a few pinnæ, large, broad, and oval, like the leaves of the tobacco-plant, spring from a creeping root-stem; while the fertile fronds, separate as in the former cases, and clothed with seed, stand erect and crowded.

From the crevices of the loose stone fences, from the angles of the buttresses of the buildings, and from the decaying walls of such as are fast going to ruin,— a painfully frequent sight in this land,—a multitude of species, reminding us in their elegantly arching habit, and their minutely divided outline, of our own *Lastreas* and *Polystichums*, spring out. *Aspidium patens* is here, with its long, slender, deeply-cut pinnæ; and *Blechnum occidentale*, with its narrow band of deep brown sori running down the middle of each leaflet; and *Pteris grandifolia*, whose pinnate arches, of ten feet in curve, transmit the tender green light of the sun through their substance overhead, making the marginal lines of opaque sori the more conspicuous; and *Polypodium Paradisœ*, a lovely form, whose multitudinous narrow leaflets, symmetri-cally parallel, are studded each with its double row of pale yellow beads; and *P. effusum*, whose leaves re-semble triangular feathers of a lively green hue; and *Asplenium præmorsum*, every part of which is narrow and linear;—and, lo! at length we stumble on our own familiar Brake. Can it be? Can *Pteris aquilina* be a denizen of these tropic plains? Yes; it grows

all over the world. The Australian savage eats its
spongy root; the elephant crushes it with his massive
foot in Ceylon; the axis deer hides in its bowery
shadow in Nepaul; the ostrich lays its great eggs
upon its *débris* in Nubia, as did the Dodo of old in
the Mauritius; the Indian in North America sleeps
on its gathered fronds, as does his brother in Brazil,
and his cousin in the isles of the Pacific; so that we
need not wonder to find its well-known visage here.

At a somewhat higher elevation we find *Polypodium lachnopodium*, or the shaggy-footed; a handsome species with broad, triangular leaves minutely
divided, and densely clothed with a red down of close,
hair-like scales; and *P. divergens*, one of noble dimensions, wide-spreading, and four times cut. Some fine
Pterides also grow here, such as *P. hirsuta*, exceedingly
rough with red down on every part; and *P. sulcata*,
with an outline not unlike that of our own *filix mas*.
Then we see *Lastrea villosa*, whose fronds, four yards
long, block up the narrow lane, as they spring from
their stout stem; and *Elaphoglossum crassinerve*, like a
massive leathery hart's-tongue, crumpled at the margin,
but with the soriferous fronds wholly brown beneath,
and with a root which creeps over the mossy stones.

Where the path winds round the base of the limestone cliffs, we find species of another character.
Here, out of the narrow clefts and chinks, project the
delicate *Gymnogrammas*, some of which display that
peculiar furniture of farinose powder that gives so
much attractiveness to their delicate fronds, and for

which we call them Gold and Silver Ferns. *G. chryso-phyllum* and *G. sulphureum* * both grow on these rocks in profusion; whose leaves on their inferior surfaces are densely covered with the bright yellow dust,—the former rather of a richer, more golden hue, the latter paler, as their specific names intimate. And scarcely less abundant is that Silver Fern, perhaps still more lovely than either, as it certainly is of nobler dimensions, *G. calomelanos*, a most graceful and elegantly-cut species, clad beneath with the purest white, to the minutest divisions of the filagree fronds. Other species, as *G. trifoliatum* and *G. rufum*, destitute of powder, and though elegant, yet less so than these, the latter, indeed, simply pinnate, mingle with them; and more rarely, examples of another genus, not less celebrated for their gold and silver clothing,—the *Nothochlænas*. Here are at least two species, *N. nivea*, and *N. trichomanoides*; the former one of the very loveliest of the Silver Ferns, from the purity of the whiteness, and from the circumstance that the dark sori are collected into a marginal belt, thus setting off the snowy area, instead of being scattered over the surface, as in the *Gymnogramma*. The latter species has long, pendulous fronds, simply pinnate, not unlike our own Trichomanes Spleenwort in form, but well silvered beneath; the whole plant the more singular, because clothed with a dense reddish down.

* It is quite common, even in works of authority, to see names ending in *gramma* treated as if they were feminine, as they appear to be: this form, however, is neuter.

Contrasting with these in their light and fragile beauty, we see the rigid, massive, hart's-tongue-shaped leaves of *Campyloneurum nitidum*, shining dark-green above, and studded with round spots of golden fruit beneath, whose scaly root-stock creeps about the base of the rocks, clinging firmly to the angles and points of the limestone.

In the gloomier part of the pass,—for the cliffs are now precipitous on either side,—we see noble tufts of a Fern, which we should say is our own Sea Spleenwort, but for the great size of the fronds, which are two feet long. And possibly, after Mr Foot's Irish experience, it may be the same; though botanists have been accustomed to call this Caribbean form *Asplenium lœtum*. Very unfamiliar, however, is the appearance of *Anemia phyllitidis*, which rears its flowering spikes at the base of these precipices; for, from a dwarf erect stem, leaves like those of the ash are thrown out, each of which sends off from the origin of the first pair of leaflets a pair of erect, slender stalks, which are each crowned with a brown spike of fructification, resembling that of our own *Osmunda*, but of finer texture. And still more uncouth, more bizarre, is yet another denizen of these fissured rocks, the *Dictyoglossum crinitum*, perhaps the oddest example of this order known. Each leaf is a roundish or elliptical lamina, some twelve inches in length, and half as broad; thick, and stiff, filled with veins united in the most elaborate network; the whole surface studded with harsh, rigid, black hairs, most abundant on the margins and on the footstalk and mid-rib, the stalks, indeed, resembling some hairy caterpillars. The leaves of this most remarkable Fern shoot up from a thick, scaly, matted root, which lines the interior of the rocky crevices.

As we approach the summit of the mountain range, where the air has a perceptible coolness, even under the beams of the vertical sun, we observe the Ferns assume a more and more prominent place in the characteristic vegetation. We cannot attempt to notice in detail a tithe of the species; look where we will, almost, we see some kind or other. The narrow bridle-paths are fringed with Ferns: out of the mossy edges that are saturated with moisture like a sponge, the luxuriant fronds spring in multiform beauty, and curve over and hide the footway for yards and yards together; while minuter forms revel beneath their protecting shades. Here is the *Campyloneurum phyllitidis*, long and lance-like, sprinkled with yellow dots, and the *Goniopteris crenata*, with its prettily vandyked margins, and the curious one-sided *Campteria biaurita*, with each of its lowest pairs of large and boldly-notched pinnæ furnished with two strong pinnules on one side, and nothing to correspond to them on the other. Here is the *Lastrea pubescens*, broadly triangular; and the *Polystichum mucronatum*, so narrowly pinnate as to be strap-shaped in outline, with curved rhomboidal leaflets, bristling with fine points, something like our Scottish Holly-fern.

The rough stones by the path-side, and the great roots of the hoary trees, which project their contorted folds like giant serpents across the choked way, are sprawled over by the lithe, and slender, and wire-like, or the thick, and massive, and gnarled rhizomes of the creeping species. *Olfersia cervina*—a handsome kind, with very distinct forms of frond: the barren, singly pinnate, smooth and green; the fertile bipinnate, very slender, and uniformly rust-red with the confluent *sori*—is one of these. And *Phlebodium aureum*, broadly pinnate, glaucous green, a noble form, throws its thick root-stock about in irregular contortions, all covered with golden hair, that shines like silk; while its dark brown rootlets, as delicately fine as threads, cling to the rough gray stone, meandering over it like a spider's web. The black, scaly rhizome of *Goniophlebium dissimile* creeps rapidly up to the summit of a tall block of stone, and allows to droop on every side its long, weeping fronds, soft, thin, and flaccid as tissue paper, but crisped, and of a fair yellow-green hue, to a distance of three feet from the base: while from beneath their protection peep out the elegant leaves of *Phegopteris hastæfolia*, spindle-shaped in outline, owing to the regular diminution, both above and below, of the leaflets, which individually are marked by singular ear-like projections at their bases. Another creeping root is that of *Nevrodium lanceolatum*, which sends off at intervals narrow, smooth leaves, of a rich, light green, much attenuated below, and at the tip abruptly contracted in a curious fashion, as if the margin there had been rolled on itself to form the golden fructification.

This region is specially the home of the most
charming genus of the whole order, *filicum facile
princeps*,—the Maiden-hair, *Adiantum*. The pretty
and delicate *A. concinnum*, resembling, and even
almost rivalling, our own *capillus Veneris*, droops its
pellucid fronds, tinged with pale red, from the rocks;
and the sweet Venus-hair itself is a native of the
island, though confined, I believe, to its sea-side
caves, from whose dripping walls, as in that one at
Pedro Bluff, accessible only at certain states of the
tide, it hangs over the mouldering bones of Indian
aborigines, who found a last refuge from tyranny in
those dreary retreats. *A. tenerum*, a Fern still more
closely like it, but with smaller pinnæ and a wider
frond, is a common form in these dewy mountain
glades; and we find in special abundance *A. striatum*,
growing in rich green tufts from every crack and
cleft, and from the rough bark of the old trees, the
fronds displaying their long, taper pinnæ, closely
studded with their multitudinous pinnules symmetri-
cally crowded.

Nor are wanting other kinds, of greater majesty, if
of inferior grace. *A. intermedium* has a fine bold frond,
beset with many pinnules of long, pointed form, which
bear the dark sori conspicuous along the edges; *A. obli-
quum* is somewhat like it, but simply pinnate; and *A.
lucidum*, whose leaflets, dark green and richly polished,
are drawn out to longer points. *A. trapeziforme* is a
noble yet very elegant species, rearing its wide but
loose leaves to the height of eighteen inches over the

wayside stones ; the large triangular or quadrángular pinnules, of a tender, light green, overlapping each other, and diminishing in size regularly to the tip, and the tall rachis, with its angled ramifications of polished ebony, so firm, yet so light and graceful. Finer still is the magnificent *A. macrophyllum*, with leaflets of similar appearance, but much larger, and more symmetrically triangular, in like manner elevated on tall, slender, erect stalks of polished ebony, which rise from a wiry creeping root-stock that insinuates itself into the cracks and hollows of the rough and moss-covered stones. And yet once more, in drier and more open spots, we discover *A. Wilsoni*, a Fern whose pinnæ, shaped like those of the last-named, are even larger yet, and produced to a long point. There are but one or two pairs of these leaflets, and a very large terminal one, the pinnæ here corresponding in form and appearance to the pinnules in the former species, which simplicity gives a very unique character to this fine Fern.

We may see also a species here, which, from the rhomboidal shape of its pinnæ, and their boldly-notched edges, together with the slender, black, erect footstalks, we might readily mistake for an *Adiantum :* it is, however, really one of the Spleenworts,—though surely a very singular one,—*Asplenium zamiæfolium*, as may be discovered by the brown fructification running in fine, nearly-parallel lines, obliquely across the inferior surface of the leaflets.

How grand are the giant trees of these primeval

forests ! The cotton-trees, the figs, the Santa-Marias,
the parrot-berries, the broad-leafs, the mahoganies,
the locusts, the tropic birches, and multitudes of
others, lift their crowns of foliage to the sky, at the
summit of hoary columnar trunks of colossal altitude
and thickness; and these massive pillars are the
homes of ten thousand parasitic plants. Some of
these trees may have a bark of almost unbroken
smoothness; others are deeply cleft and fissured; but
the minute seeds of the parasites find means to effect
a lodgment in each, and the growing roots cling
firmly to the surface. From the perpendicular trunks,
from the hollows and forks of the greater ramifications, and all along the surface of the huge, horizontal
limbs,—each of them a forest-tree for dimensions,—
spring great tangled tufts of orchids and wild pines
and ferns, and climbing bignoniæ of gorgeous bloom,
and long, depending lianes, some as thin and pliant
as whipcord, others woody, thick, and twisted, like
huge cables; making the penetration of the woods
difficult beyond conception.

Of the glorious beauty of the flowering species I
must not here speak; we have enough to do to mind
the Ferns. There is one springing from the bark of a
fig-tree, which you would take for a loose tuft of
grass, so long and grass-like are the narrow leaves.
But no; it is a real Fern, *Campyloneurum angusti-
folium,*—one of the great *Polypodium* genus; as is
seen by the round red seed-groups running in a double
line along the under side. Here is another, hand-

somer certainly, if less abnormal, *C. reptans :* it is one
of those which have a hart's-tongue outline, only more
elegantly pointed; and having the large conspicuous
sori running in parallel diagonal bands, while the
deep green of the upper face is dotted over with white
scales. These leaves, like those of our own polypody,
are jointed to a scaly root-stock, which clings to and
creeps over the limbs of the trees. And yonder we
discern one which we might readily mistake for that
familiar tenant of our hedges, a little more luxuriant
in dimensions, and a little more pendent: the creeping root beautifully speckled with brown or light
green, like the body of a snake, gives it a character,
however, of its own. It is *Polypodium loriceum.*

From a matted mass of heterogeneous foliage
filling the broad fork of two vast cognate limbs in
this Santa-Maria tree,—a mass in which you might
easily distinguish a score of species crowded together,
—stands up a stout bundle of united stems, from
whose summit diverges a crown of short, dark-green,
oblong leaves. Beneath they display slanting lines
of fructification, set in graduated series, the full-
length, half-length, quarter-length, and one or two
shorter yet, alternating; while little leaves are forming in a proliferous fashion at the bases of the old.
This is *Diplazium plantagineum,* a curious example
of the Spleenwort family. And here we see a still
more remarkable member of the same household,
seated on the limb of another tree. The leaves a
simple oblong, the fertile ones blunt, the others

pointed, all only a few inches in length; the former
remarkable for the large size of the very conspicuous
horse-shoe-shaped sori, which, towards the point of
the leaf, overlap one another. The taper extremity
of the barren leaves gives origin to other leaves,
which add to the odd appearance of this little Fern.
It has been named *Fadyenia prolifera*, by Sir William Hooker, in honour of the late Dr James Mac-
fadyen, of Kingston,—a most worthy man and excellent botanist, whose admirable work on the botany
of Jamaica was interrupted by his lamented death.

Besides these, which are stationary and local, there
are other Ferns of a much more restless habit, which
creep rapidly about, and, roaming over and around
these gigantic trunks, seem to claim them for their
own throughout their vast length. *Polypodium vac-
ciniifolium*, from an irregularly meandering root-
stock of a yellowish hue, densely covered with shaggy,
pointed scales, which clings to the bark, and crawling
up and up, like a rough caterpillar indefinitely extending itself, puts forth at short intervals a number
of small heart-shaped leaves, that look like those of
some evergreen phenogamous plant, rather than the
fronds of a fern. Then there is *Goniophlebium
piloselloides*, which has a similar habit, with a much
slenderer clinging stem, not thicker than a bit of
copper wire, which it much resembles, sending forth
its hair-like roots on each side, over the smooth, gray
bark, and giving birth to stiff leaves clad with short
hairs. These differ; the barren ones are oval; the

fertile ones nearly twice as long, and only half as broad, with two rows of very large, round sori beneath, like strings of golden beads. *Phlebodium lycopodioides*, again, on a root-stock of the hairy caterpillar type, has leaves elegantly spindle-shaped, and the seed-beads separated,—a different but equally attractive species; while *Oleandra nodosa*, with a similar root-stock, red and shaggy, has pointed leaves closely resembling, in size, form, and colour, those of the elegant oleander, and the seed-masses scattered in minute dots.

These all have entire fronds; but there are also the species belonging to the beautiful genus *Nephrolepis*, which have the climbing habit, whose fronds, long and narrow, have the pinnæ set at right angles, like the teeth of a comb; as *N. exaltata*, for instance, whose lovely green leaves, extending to a length of four or even five feet, with a width of three inches, carry each a hundred and twenty pairs of leaflets, set in the most perfect symmetry. *N. pectinata* has still narrower leaves, and more numerous pinnæ. Viewed from beneath, as the very elegant fronds arch out from the supporting tree, these ferns have a singular appearance, for the mid-rib is quite concealed by the bases of the pinnæ; and these having each a sort of ear-like projection on the anterior side, the alternation of these swellings, wrapping over the central rib, imparts a curious waved form to the dividing line; while the double lines of seed-dots on the pinnæ beyond the middle of the series greatly augment the elegance of these very lovely Ferns.

Among the climbers we have yet another, and a quite diverse, type of Ferns, the genus *Davallia*, noted for their large, broadly-triangular, minutely-divided fronds, and for the sori being placed, like very minute black pin-heads, at the tips of the lacerations. *D. elegans* runs up tree-trunks as well as over rocks, in these humid mountain woods; its stout, pale-brown, woolly root-stock, conspicuous as it winds, connecting the somewhat remote but magnificent fronds, which are two feet in length, and almost as much in width. It is a species of great elegance and noble beauty.

In the very heart of the tangled thickets, where the earth is soft, and black, and spongy, where a carpet of lovely green *Selaginella* is spread in great sheets over the roughnesses of uncouth stones and old decaying logs, and where the light of heaven finds its way softened and subdued through myriads of twinkling green leaves far overhead, many kinds of Ferns delight to grow and flourish, as the vivid hue and freshness of their tender fronds reveal. Here we may find *Diplazium striatum*, an elegant array of large but exquisitely delicate pointed leaves, spreading from the summit of a slender stem, a foot high,—a Tree-fern in miniature. Or we might suddenly break upon the magnificent *Hemidictyon marginatum*, with its broad, translucent pinnæ permeated by a charming network of veinlets towards their outer edges, and marked by strong oblique lines of brown sori,—one of the noblest of non-arborescent Ferns, sometimes shadowing an area of five-and-twenty feet. Or we might meet with an-

other subarborescent species in these dense matted
woods, *Marattia alata,* whose fronds have the peculiarity of springing out of the thick, fleshy stem, between
two appendages which resemble abortive fronds, and
whose fructification is entirely destitute of the elastic
ring, which is so characteristic of this great tribe of
plants, raising this genus to a higher rank than other
Ferns, and presenting the closest approach of any to
the flowering plants. Among Ferns of humbler pretensions, the pretty little Bird-foot, *Cheilanthes radiata,*
spreads its maidenhair-like pinnæ in the form of five-
rayed stars, or like slender-armed star-fishes changed
to green plants by one of Ovid's metamorphoses, each
at the summit of a tall, erect, slender stalk. And
beneath the shadow of the lofty *Marattia* reclines the
curious Ivy-fern, *Hemionitis palmata,* whose five-
angled leaves, grovelling on the ground, clothed with
a bristling crop of red down, scarcely look like those
of a Fern at all, till you gather one, and hold it up to
the light, when the network of veins appears; unless,
indeed, the plant is in fructification, when the fertile
fronds, standing erect, display on their inferior surface
the rust-brown lines of sori exactly corresponding to
the veins, an exquisite net of dark red meshes spread
over the pale green leaf.

Should our way lead us into one of those deep narrow gullies or gorges which so frequently occur in the
limestone formations of the Jamaica mountain-region,
we might see Ferns of yet another type, such as occur
only in such or similar conditions. A rank, coarse

vegetation conceals the wet ground, among which the
dangerous dumbcane towers, with its broad arrowy
leaves, of whose juice a single drop is sufficient to
swell up the lips and tongue, and preclude speech.
The hard gray limestone rises in steep walls nearly
meeting overhead, all fretted and eroded with deep
hollows and sharp points, like the rocks of our Devonshire coasts. In these little cavities elegantly twisted
shell-snails reside; and in the larger ones, always brim-
ming with water, the shrieking tree-toads delight to sit,
enjoying their cool bath. And out of the same crevices many species of Film-ferns, *Hymenophyllum* and
Trichomanes, project their tufts of pellucid fronds,
and twine their matted wiry roots around the groined
projections. In the same fissures, and out of the
rough bark of the tall trees that rear themselves
towards the light by the wall-like cliffs, spring several
kinds of *Gleichenia*, (*G. Bancroftii*, and others ;) a
genus of Ferns of singular aspect, possessing wide-
spread fronds of very lax habit, and of very minute
segments, but so peculiarly elegant and delicate, that
they have been termed the aristocracy of the fern-
tribes.

Last, but not least, we emerge on a region where
the true arborescent species, the tall Tree-ferns, grow
in their majesty.

The handsome *Dicksonia cicutaria*, when old,
forms an umbrella of vivid glossy green fronds, set
on a true tree-like stem of considerable thickness,
though of no great height. And *Hemitelia grandi-*

folia, a Fern of very different aspect, is also of low altitude, rarely exceeding the height of man: its broad, pointed pinnæ, cut into knife-like teeth, give it a peculiarly noble appearance. *H. horrida*, however, attains a really arboreal height; a species whose stem and mid-ribs bristle with sharp spines, and whose young leaves have a remarkable appearance, from being clothed with a sort of gray wool on their first unfolding.

Cyathea arborea is a species of peculiar elegance, growing in more open spots, in small groups and groves. The slender stems, each marked with its oval scale-like scars, and throwing out from its summit its swelling cluster of leaf-bases, so compact and so regular as to look like the elegantly-fluted knob of some cast-iron pillar, again constricted before they spread abroad in a wide umbrella of finely cut foliage, have an imposing effect, surrounded by the moss-grown trunks, shaggy with gorgeous parasites, of tall trees that tower up and interweave their branches far overhead, shutting out the sun, and almost the light.

The *Alsophilæ*, again, are Tree-ferns of lowlier elevation. *A. ferox*, rightly named, since it is most ferociously armed with long, rigid thorns, rears a stem three or four yards high, from the midst of rank herbage. Its fronds are generally like those of our own male fern, but exaggerated: the formidable prickles, however, that stand up from the knobbed bases of the fronds, which swell out around the summit of the trunk, like the bulging branches of a

z

candlestick; the elongated scars on the stem, that
mark the position once borne by the now-fallen fronds;
and especially the lower half of the stem, so clothed
with roots as to look like a mass of intertwining wire,
black and shining, and running down with the concentrated moisture of these damp woods,—are totally
unlike anything ever seen in a temperate climate.
Finally, there is another species of the same genus
abundant in these lofty woods, *A. pruinata*, which,
had it an altitude commensurate with its expansion,
would be one of the most magnificent ferns in the
world. Instead of spines, its trunk is invested with
woolly hair; and its minutely divided foliage, elegantly tapering, and of a tender green above, is
covered beneath with a silver hoar, like that of the
Gymnogrammas, in which it is the rival of the most
magnificent of all Ferns, *Cyathea dealbata*, the Silver
Tree-fern of New Zealand.

Thus our imaginary tour through one tropical
island would bring before our notice examples of
almost every important type of form included in this
immense order. Indeed, there is no notable exception
but the *Platyceria* or Stag-horn Ferns—those remarkable parasitic forms that cling to the trunks of great
trees, and have two kinds of fronds, one globose and
bent downward, the other flat and palmate and
spreading. No example of this curious Fern is found
in the western hemisphere.

Drynaria, a sub-genus of the great genus *Polypodium*, though not unrepresented in the western

hemisphere, has its home in the vast islands of the Oriental Archipelago, where it assumes some singular and interesting forms. Some of these have fronds of two kinds,—the fertile, stalked and deeply divided; and the barren, which, springing from the great woolly rhizome without any stalk, are broad, and cut at the edge into rounded sinuations, like enormous oak-leaves: these, in withering, become of a rich brown hue, but retain their form and position, and greatly augment the beauty of the specimen. Such are *D. quercifolia; D. diversifolia,* &c. Others again, as *D. morbillosa, D. coronans,* &c., have broad sinuate fronds of immense size, springing at intervals from a very thick scaly rootstock; and as this has the habit of creeping around the stout slanting limbs of giant trees in their native isles, the stalkless leaves arching gracefully outwards at their summits cause the whole plant to resemble a majestic coronet, through the centre of which the branch passes.

Few of us, however, have the opportunity of wooing Flora in her tropical bowers; we content ourselves, therefore, with gathering her treasures, and improvising the tropics at home. Fortunately, the lovely Ferns are domesticated with little difficulty; and, a few principles of culture having been mastered, we are able to grow them to a luxuriance and beauty often even superior to what they attain in their native haunts. And when wealth and taste combine, what can be more charming than the ample stove-ferneries erected and furnished by some of our princes of horti-

culture! Such a one is that at Rockville, near
Dublin, the residence of Mr Thomas Bewley. It is
a house of about sixty feet square, divided into three
aisles by rows of rustic arches and pillars, the centre
being twenty feet in height, the sides a little lower.
It is heated to a tropical temperature by hot-water
pipes, and is covered by a *double* glass roof, an admirable contrivance for maintaining an equable temperature, six inches of air between the two roofs acting as a non-conducting blanket. The inner glass is
stained of a wine-red colour, which imparts a wonderfully rich tone to the light within, while it subdues
it to that degree of shade most congenial to these
shadow-loving tribes.

The visitor enters through a glass door, and steps
down on its lowered floor of clean shingle. He seems
to have entered the precincts of some ancient fane,
now falling into ruin, where vegetation is silently
but rapidly exercising unchallenged dominion. The
massive buttresses of rugged gray stone, which divide
off the area on each side, and the pointed Gothic arches
that spring from them, are built with gaping joints,
and rude irregular projections, in and over which
mosses and lycopods and ferns are growing, arching
out, depending, or creeping over the coarse shaly
stone, and everywhere presenting the most charming
effects with their lovely green foliage; requiring indeed a stern pruning, to prevent the rocky supports
from being completely concealed under the confluent
verdure.

" While the buttresses are thus decorated," writes a describer, " the open spaces of the aisles are chiefly occupied by large plants with fine foliage, mostly planted, or seeming to be so, in little mounds of rockwork. The following are some of the most striking that thus placed singly cannot fail to command attention:—A noble plant of *Latania Borbonica*, with its beautifully plaited fan-like leaves ; a noble plant of *Dicksonia antarctica*, 10½ feet in height to top of fronds, fine, dark, clear stem 3 feet in circumference, and the fronds forming a circle 12 feet in diameter ; *Dicksonia squarrosa*, true, 11½ feet in height, with noble fronds, and feathered with offsets all up the stem ; *Alsophila australica*, 10 feet in height, with an elegant crown of healthy fronds ; *Alsophila excelsa*, 11 feet in height, with noble foliage. Then there were fine large plants of *Cyathea dealbata*, *Cyathea medullaris*, *Alsophila Macarthiæ*, *Rhopala De Jonghi*, and the beautiful *Aralia leptophylla* and *papyrifera :* also good plants of the singular-leaved *Dammaras* from New Caledonia ; *Dracæna nutans ; Imatophyllum miniatum*, 12 feet in circumference ; a Chinese colt's-foot, *Farfugium grande*, 20 feet in circumference; and a noble specimen of the India-rubber plant, which, with its thick, leathery leaves, contrasted strongly with the feathery foliage by which it was surrounded.* . . . .

" The north end of the house is a solid wall, and in front of it is a fine massive, irregular specimen of

* These last-named plants are not ferns.

rock-work formed of different materials, but each by itself, and thus on a limited scale furnishing materials for geological study. These, so far as we recollect, were Ballycoras clay slate, and granite, gray granite, quartz, red sandstone, conglomerate, tufa, petrified moss, &c. The lower part of the pile is not only irregular, but formed into arched vaults, caves, recesses, nooks, and crannies, to suit some sweet little things that modestly like retirement from the glare of bright sunlight; as the varieties of Killarney fern, and some other of the *Trichomanes*, and such *Hymenophyllums* as *Tunbridgense* and *Wilsoni*, and other small ferns and lycopods. Then on the face of the rock were fine specimens of *Platyceriums*, good foliage of *Begonias ;* and among other ferns and mosses, we have a vivid recollection of a *Platyloma* throwing its elegant fronds over red granite. Steps, rude as they ought to be, take you from either side over the top of this rock-work, revealing some rarity and beauty at your feet at every step, until, reaching the top, and surveying the whole,—the wreathed buttresses, the draped arches, and the expanse of the fine foliage of tree-ferns, &c., beneath your eye—you might easily imagine you were standing amid the ruins of the buildings of a forgotten race, such as are to be found in Central America, where vegetation in wild melancholy grandeur is revelling amid, and obliterating, the evidences of a previous power, genius, and civilization."*

* *Journal of Horticulture*, March 25, 1862.

In such a retreat as this the amateur pteridologist may watch the gradual development of his lovely favourites, admire their manifold beauties and graces, and accept the smiles of gratitude with which they greet him, as they root into the nooks and crannies, or comfortable pots prepared for them, all well stored with the fibry peat in which they delight, or the new-fashioned cocoa-dust, which seems to suit their appe-tites better than anything else; and here, if he be a Christian, he may lift his heart in loving adoration and praise to the blessed God, who has adorned the earth with such loveliness for His own glory. And, surrounded by such opportunities, the botanist may most advantageously pursue those investigations on the structure, the germination, the growth, and the fructification of this tribe of plants, which are yield-ing results so remarkable, so striking. Some of these results, as being the most interesting facts connected with the history of ferns, I will endeavour briefly to describe, under the guidance of that able botanist, Dr Wilhelm Hofmeister.

How does a Fern reproduce its kind? Seeds, such as those which flowering plants produce, it has none; but it yields millions of organs which serve instead of seeds to originate a new generation of its kind, though their structure and the mode of their development are totally diverse from those of seeds. If we take a leaf of a *Polypodium*,—the common *P. vulgare* from a hedge will do just as well as any exotic species,—we shall find, if it is in fruit, that its back is studded

with a number of round spots called *sori*, very conspicuous from their golden yellow or rich brown hue.
Examination with a powerful lens reveals that each
of these spots is a group of tiny globules, known as
thecæ, each elevated on a slender stalk, and closely
crowded. A microscope is, however, needed to discern
more.

By the aid of this instrument we perceive that each
theca is a hollow sac, strengthened by a stout *ring*,
marked by transverse bars, that passes vertically round
it, like the brass meridian round a globe. At a certain period this elastic ring bursts, rupturing the walls
of the *theca*, and scattering its contents, viz., the *spores*,
which are the ultimate agents of reproduction, answering to seeds. They are, however, homogeneous corpuscles, generally of an ovate or rounded form, with
three more or less prominent ridges running along
them longitudinally.

Under the influence of warmth and moisture, the
thick but brittle outer coat of the spore bursts, and
the contents protrude as a clear bladder, which by
growing assumes a tubular form, divided by transverse
partitions; in other words, it is a single linear series
of oblong cells, within which, on the inner surface of
their walls, grains of green substance, (*chlorophyll*,)
the colouring matter of plants, develop themselves.
At the same time, a slender root grows out from the
lowest cell. After the conferva-like filament has
made its fourth or fifth linear cell, the terminal one
divides longitudinally, so that two oblong cells placed

side by side now end the series. These now go on developing other cells, by division both transverse and longitudinal; the result of which is, that the tip of the confervoid rapidly assumes a flat, fan-like shape, and is recognisable as the *prothallium ;* a condition in which the new growth very closely resembles the common Liverwort (*Marchantia*), which spreads its expansions so commonly over the earth in damp situations.

We shall be greatly mistaken, however, if we suppose these green laminæ which lie so thickly overlapping each other, and sparkle so prettily, in our seed-pan, to be the future Ferns. The *prothallium* does not enter into the composition of the future plant at all; it is but a sort of foster-mother by which it is reared. In it, however, occur a series of developments of most remarkable character. When the *prothallium* has attained dimensions which make it distinctly visible to the naked eye, minute warts begin to form on its under surface, which are called the *antheridia,* from their performing functions analogous to the male organs (*anthers*) of higher plants. Each *antheridium* is composed of a large central cell, supported by one cylindrical or two semi-cylindrical cells, covered by a cell having the form of a segment of a sphere, and surrounded by a ring of several smaller cells. By and by the central cell, having increased in size, is transformed by internal divisions into a globe containing several cubical cells; and in each of these latter there is now produced what has the form,

and appearance, and motions of a living animalcule;
a flat ribbon-like worm, spirally coiled into about four
whirls, tapering to a fine-drawn point behind, and
furnished for about half its length anteriorly with a
number of projecting hairs (*cilia*). At first this
worm (*spermatozoon*) lies coiled motionless in a little
transparent bladder, within one of the cubic cells.
The walls of these cubic cells now dissolve; and the
globules with their (as yet inactive) *spermatozoa* lie
loose in the midst of the *antheridium*. At length the
terminal lens-shaped cell of this bursts, and the globules escape, and swim with a rotary motion in the
drops of moisture which lie condensed on the inferior
surface of the *prothallium*, the ciliated end of the
spermatozoon protruding through each. Suddenly
the globule bursts with a wide aperture, and the *spermatozoon*, partly uncoiling, darts out and swims away
with a rapid motion, rotating as it goes. But we
must leave these curious bodies awhile, and trace the
development of another equally strange set of organs.

The *prothallium*, meanwhile, has been growing, and
has taken a forked form, the tip forming two expanded
portions, divided by a deep indentation. The bottom
of this sinus is the seat of the *archegonia*, which are
but few, while the *antheridia* are very numerous.
The substance of the *prothallium* on its under-surface,
just behind the bottom of the sinus, becomes a cushion
of small cells, by the minute subdivision of the cells
already existing at that part; and on this cushion
there grows an elevated wart, which is the *arche-*

gonium. Within its base it holds a globose cell, which becomes the embryo-sac, while the elevated portion becomes a sort of chimney or shaft, composed of four courses of four cells each, the top of which is as yet closed. The embryo-sac encloses a single nucleus at first; but presently a second is developed, which becomes the germ-vesicle; it is minute at first, but rapidly increases, while the primary nucleus shrivels and disappears. At length, the cells which close the summit of the shaft burst apart, and its fluid contents escape, leaving an open passage to the embryo sac. At the same time, the wall of the latter softens and dissolves, and the germ lies exposed at the bottom of the open shaft.

Let us now return to the *spermatozoa,* which are whirling about in giddy gyrations in those minute drops of condensed moisture, which lie studding the surface of the *prothallium,* like seed-pearls. Often these drops coalesce; and the thin space between the *prothallium* and the earth is a continuous stratum of water. One of the *spermatozoa,* finding the summit of an *archegonium* open, instantly enters, and makes its way to the bottom, where the germ lies, around which it plays sportively. This is impregnation: as soon as it occurs, the end of these wonderful provisions is attained; and the shaft immediately closes again by the swelling of the terminal cells. The germ further enlarges, and divides into four cells in one plane, one of which grows into the bud and first frond of the young fern, while another produces the root. The active growth of these, respectively upward and downward, produces an ever-increasing expansion of the surrounding tissue of the *prothallium*, till the latter is at length ruptured, and the young frond protrudes, curves upward, and appears between the two flaps of the *prothallium*. Before this it has formed the lamina, which is always much less divided than the mature frond. The root also protrudes downward, and penetrates the ground. Such are the marvellous processes which attend the earliest life of these charming plants. The record reads like a fairy tale; and, but for the numerous witnesses—witnesses of the highest scientific acumen, and the most unimpeachable veracity—who confirm the testimony, it would be dismissed as a myth. The facts, however, are beyond dispute; and, indeed, may without much difficulty be verified by any one accustomed to the more delicate microscopic investigations. The following details by Dr Hofmeister will teach the student what to look for, and how :—

" When a quantity of fern-spores are sown, the germinating *prothallia* are developed at very different periods. The earliest *prothallia* produce in the first instance only *antheridia*, afterwards *antheridia* and *archegonia* together, and when advanced in age, only *archegonia*. The earliest *prothallia* have already attained the latter stage at the time when the later *prothallia*, the development of which has been retarded by the shade afforded by the earlier ones, are thickly covered with *antheridia*. If the plants are now kept for some days rather dry, and then saturated with water, the result will be, that numbers of *antheridia* will emit *spermatozoa*, and numbers of *archegonia* will open contemporaneously. The water should not be poured over the plants, but the pot should be placed in water nearly up to the margin, by which means capillary attraction and condensation will yield abundance of moisture to the *prothallia*. After one or two hours, the surfaces of the larger *prothallia*, which are covered with *archegonia*, are found almost covered with *spermatozoa*, partly in motion, and partly at rest. If a delicate longitudinal section through the parenchyma of these *prothallia* be examined immediately, with a magnifying power of from two hundred to three hundred diameters, *spermatozoa* are sometimes found in all the *archegonia* along the whole length of the section. I thus found three *spermatozoa* in active motion in the central cell of the *archegonium* of *Aspidium filix mas*. In this case the motion ceased seven minutes after the commencement of the observation, and was accompanied (probably caused) by the coagulation of the albuminous matter of the cell-contents. In the same fern on two occasions, and also in *Gymnogramma calomelanos* and *Pteris aquilina*, I have seen a *spermatozoon* in motion in the central cell of the *archegonium ;* and in the above-mentioned species, and also in *Asplenium septentrionale*, and *filix fœmina*, I have seen a motionless body near the germinal vesicle, (after the growth of the latter has commenced,) answering in form to a *spermatozoon*. Lastly, in *Aspidium filix mas* and

Pteris aquilina, I have often seen motile *spermatozoa* in the canal of the opened *archegonia*, the motion of which *spermatozoa* ceased during the continuance of my observations. I may add, that these observations were very numerous, and were undertaken with the view of following out the cell-development of the embryo. In a single *prothallium*, cultivated in the manner stated above, and laid open longitudinally as I have mentioned, there will not be found more than three, or at the most four *archegonia* open at the apex; *spermatozoa* will probably be found in not more than one in thirty of such *archegonia*, and they will often not be found at all."*

The increase of specimens by means of buds or bulbules must not be confounded with germination. By germination a new generation is produced; the little fern that grows out of the *prothallium* being the daughter of the plant, whose spore produced that *prothallium*. The developed bud, however, is but an essential portion of the individual fern on which it grew, and partakes of any accidental (*i. e.*, not specific) peculiarity possessed by it. The production of such adventitious buds is, however, a highly interesting phenomenon to the fern-cultivator; especially when they occur upon the leafy part of the frond, as in many species of *Asplenium*. If we examine one of these proliferous species, *A. odontites* or *A. viviparum*, for example, we shall see young plants in every stage dotted about on the surface of the fronds,

* Hofmeister, *Higher Cryptog.*, p. 198, note.

from the minute black scaly wart which breaks out
of the membrane, to the well-formed fern with four
or five fronds. The leaves on which this phenomenon
occurs are flaccid and procumbent; and as they gradually decay on the moist earth, the new plantlets
strike their roots into the soil, and become independent. *Woodwardia radicans,* again, a noble species
from the south of Europe, forms, near the point of its
fronds, a plant, which grows to such an extent as to
have, not rarely, half-a-dozen fronds a foot in length,
yet deriving all its support from the main plant.
Others, as *Aspl. radicans, A. rachirhizon Adiantum
caudatum,* &c., have the tip of the frond drawn out
into a long filament on which the young plants grow:
and it is a common mode of increasing such species
to peg the tail down to the earth with a hair-pin,
where the progeny soon root, and can be potted off.
Again, *Cystopteris bulbifera* forms little bulbs on the
under side of the mid-rib. These are green, egg-
shaped bodies, nearly as big as peas, composed of two
very thick fleshy leaves, like cotyledons, joined to the
rib by a connexion which severs with a slight touch.
They fall on the ground; or they may be placed upright on the soil of a pot, when in a few weeks they
send out tiny fronds of the proper form, and a plant
is made. There is a remarkable mode of increase in
the genus *Nephrolepis.* These elegant Ferns, known
by their lengthened comb-like fronds, send forth beneath the soil long runners like slender wires, which
at their extremities develop a thick oblong knob. On

the surface of this a number of buds grow out which
are prolonged into young plants, and then the knob
and wire decay.

Many persons must have admired the noble specimens of *Platycerium alcicorne*, which for several years
has grown suspended against the wall of the tropical
fernery in Kew Gardens. But perhaps few have
penetrated the mystery of those great succulent semi-
globular leaves, like those of a large cabbage-heart,
which adhere by their edges to the suspended board.
The first frond formed is upright and spoon-shaped;
then one comes, which is thick and circular, projecting horizontally, and presently curving downward:
after this, erect fronds again grow, which develop
flat forking tips, like the horns of an elk; and then a
pair of the globose recurved form, one on each side,
covering up the base and the roots in their ample
hollow. Why this peculiarity of growth? The Fern
is a native of tropical Australia, where it grows on the
perpendicular trunks of great forest trees, and must
often be exposed to droughts. These thick recurved
fronds afford a protection to the root and heart of the
plant, enclosing those parts in a tight box, from which
evaporation with difficulty proceeds. Other species
of the genus, whose fronds spread over the ground,
protecting the base of the plant and the surrounding
soil with their shadow, are not exposed to the same
peril, and are entirely devoid of this peculiar manner
of growth.

The microscopist may find many highly interesting

subjects of investigation in a well-filled stove-fernery.
This *Platycerium*, for instance, has its fronds covered
with a gray hoar, which, on being magnified, is resolved
into a vast multitude of isolated groups of short slender
white filaments, radiating from a centre in a star-like
manner, from six to ten threads to each star. The
gold and silver ferns, too, already mentioned, afford
pleasing objects. The very lovely little *Nothochlæna
flavens* I have just been examining ; and most charming it is. Under a power of 100 diameters, with
light reflected from a lieberkuhn, the under side of a
pinna exhibits an area which we might suppose sown
with flower of sulphur, which, however, has fallen in
tiny coherent masses, as if slightly damp, uniform in
size, and no thicker than a single layer. Over this
area the globular spore-cases are spread, in two bands,
parallel with the edges of the pinna, and running
from base to tip; these bands, however, resolvable
into oblique lines of *thecæ*, which follow the course of
the veins. They look like marbles, or rather bullets,
in which the seam of the mould may represent the
stout ring, which passes vertically over and round
each *theca;* the colour, a deep brown almost black,
with a rich warm hue appearing between the nodules
of the ring. The yellow dust, when carefully scraped
off with a needle-point, and spread on a slip of glass,
is found to be composed of loosely-adherent granules
of irregularly-ovate form, so minute that even with
very high powers, as 600 or 800 diameters, they
appear only as points; and, when measured with a

2 A

delicate micrometer, are found to average not more than 1-20,000th of an inch in diameter, while many are as low as 1-50,000th. In *G. chrysophyllum* the appearance of the gold-coloured dust is in general the same; but it is arranged more distinctly in corpuscles, which stand up from the surface of the frond on slender foot-stalks; the whole not unlike tiny cauliflowers. Both heads and stalks are pellucid and colourless, but are dusted all over, the heads most abundantly, with fine sulphur-like granules, which also lie scattered thinly over the ground between the stalks. When scraped off with a needle, the granular texture under a high power is scarcely manifest, the substance forming coherent lumps, somewhat like a very crumbly cheese. *G. Martensii* shows stalked processes as the last; but under a power of 600 diameters these are less coherent, and composed, not of granules, but of an infinity of slender short rods, cohering in every possible direction, about 1-20,000th of an inch in thickness, and of varying length, straight or variously curved, and translucent in texture. *G. Tartareum*, the best of the silver ferns, is very white beneath. Stalked processes are discernible, but few, and almost concealed by a profusion of downy silvery spicules, looking like the finest hoar-frost on a blade of grass. Under a power of 600 diameters, we find fibrous transparent rods as before, but with much less coherence, and resembling a spicular sponge after passing through fire, or the deposit which remains after a good deal of cutting of cloth, roughly brushed into masses. What is the mode of origination of these farinose granules, or what the functions they perform in the economy of those species which possess them, if they have any beyond mere ornament, has not, so far as I know, been determined.

Much more that is curious and interesting might be noted in the history of these graceful plants; but the limits of my space compel me to mention but one point more, the phenomenon of variation. That ferns are more liable to what is technically called "sporting," than other plants, is shown by the great numbers of diversities from the normal condition which are already registered. Thus, in the Catalogue of the eminent fern-growers, Stansfield & Son, with its Supplement, we have the named and described varieties of a few well-known native species advertised to the following extent:—*Athyrium filix fœmina*, 56 forms; *Asplenium adiantum-nigrum*, 29; *Blechnum spicant*, 53; *Lastrea filix-mas*, 34; *Polystichum angulare*, 43; *Scolopendrium vulgare*, 90:—making, out of these six familiar species, three hundred and five named varieties offered for sale by one firm. The possession of well-marked varieties is quite as highly valued by most amateurs, as that of distinct species; hence, there is a constant stimulus to the search for new divergencies; while the discovery that monstrosities once found are not only in general constant, but have a strong tendency to perpetuate themselves by spores, and even to originate new and stranger forms, is rapidly augmenting the stock. Mr Bridgman, in

an interesting memoir, " On the influence of the
Venation in the Reproduction of Monstrosities among
Ferns," * has recorded some very curious facts in this
direction.

" In the first instance, a leaf from the multifid
variety of the common Hart's-tongue (*Scolopendrium
vulgare*, var. *multifidum*) had been procured, selecting one of the most distorted, and the spores from it
collected indiscriminately and sown. The plants coming from these, to the extent of many hundreds, presented every grade of variation, from the simple
ligulate with a single acute apex, up to the complex
form of the parent, and beyond, or, as fern-fanciers
express it, 'greatly improving the sport;' and this not
in one direction only, but resulting in the production
of three distinct varieties. The direction of the veins
in the lower portion of the leaf from which the spores
had been taken, was all but normal, some parts
entirely so, upon which several of the sori had been
placed. But towards the upper and above the middle
portion of the leaf, the veins, losing their regularity
and parallelism, became somewhat zigzag and reticulate, the indusium only partly developed, the sori
smaller, more numerous, and nearer to the external
margin. In the extreme upper or multifid and crisp
terminal expansion, the mid-vein became broken up
into a number of nearly equal divisions, and these,
again, dividing and subdividing into a reticulate mass
of veins and venules. Instead of the regularly-formed

* *Annals and Magazine of Natural History for December*, 1861.

sori, the spore cases were distributed about in patches, without the slightest trace of an indusium, and attached by their pedicles to the back of some of the larger bundles of veins, and also in the axils, in scattered masses, the indusium having become perfectly obsolete.

" Another variety, the ' *laceratum'* of Moore, ('Nature-printed Ferns,' 8vo edition, vol. ii., pl. xcii.,) was now selected, having the two characters of venation separate and distinct. The sori from all the reticulate portions of the leaf were carefully scraped off, and the spores sown in baked peat in a pan by themselves. The plants resulting from these (which were pricked out from a seed-pan four inches in diameter, where they had come up as thick as they could grow) *contained not a single plant which had not the strongly-marked characteristic of the variety, and some far more crested and crisped than the parent.*

" The spores from the remaining part of the leaf were sown in another pan, at the same time, and have produced an equally abundant crop. *There were not a dozen plants of the same character with the preceding; and, until the leaves were several inches long, with the exception of here and there a twin-leaf, there were no external characters in the bulk of them to render their parentage recognisable.* A very large proportion of them were discarded as normal; and the only peculiarity at present shown among the remaining ones, the leaves of which average from six to nine or ten inches in length, and from one and a-

half to two and a-half inches in breadth, is in a
slightly sinuous margin, an occasional division of the
apex into two or more lobes, and a disposition to
become somewhat ragged ; and this by no means
general, but only one or two leaves on a plant.

" Similar experiments with other varieties and
species have been attended with corresponding results.
The tufted end of the variety ' *Crista galli* ' of the
same species (*Scolopendrium vulgare*) produced many
hundreds of plants, all, with scarcely an exception,
equally complex with the original, or more so ; and,
what is more remarkable, the parent-plant was upwards of two years old before it began to develop
its peculiar character, while the progeny raised from
it were all prominently characteristic in the first
leaves.

" With such forms as *Nephrodium molle corym-
biferum*, *Lastrea filix-mas cristata*, *Scolopendrium
marginatum*, &c., where the entire frond has become
deformed and the whole of the venation abnormal,
the plants raised from spores procured from any part
of the leaf reproduce the variety with little or no variation. Out of some thousands of *Filix-mas cristata*
seedlings, only one reverted to the normal form, and
two others closely approach the *angustata* of Sim, all
the remainder being identical with the parent."

At present, our recognition of this peculiarity extends
scarcely beyond British species ; though some signal
exceptions have occurred, as in the case of *Nephrodium molle*. Perhaps, however, many of what are

recorded as closely allied species among the exotic
denizens of our ferneries, may be only varieties ; and
if we had as great abundance of specimens for observation, we might find as great a diversity as among
our own. *Asplenium præmorsum*, for example, a
widespread tropical species, varies greatly in its forms
even in our stoves.

While variation of form appears thus to be peculiarly prevalent, variation of colour has been, up to a
very recent period, quite unrecognised among ferns ;
though among phenogamous plants it prevails widely.
But the new-fashioned love for piebald plants appears
to have created a supply for the demand even here.
Within a year or two four distinctly marked species
have occurred, and have been very largely multiplied ;
as, like the monstrous forms, the colour-variegation
proves constant. The first variegated fern that was
announced was Mr Veitch's *Pteris argyrea*, which
appears to be a condition of the noble *Pt. quadriaurita*
of India.   Then M. Linden, of Brussels, produced the
much more showy *Pt. aspericaulis*, var. *tricolor*, with
its fine purple mid-rib running through a broad stripe
of white.   And then *Pt. Cretica*, var. *albo-lineata*,
was sent to Kew, from the mountains of Java.   Now
our own brake is added to the list; for Mr Stansfield,
of Todmorden, informs us, that, making a botanical
excursion with Mr Eastwood lately, the latter discovered a patch of *Pt. aquilina*, "so beautifully and
distinctly variegated, that it had the appearance of
being sprinkled with snow ;" and it figures in his cata-

logue as *P. aquilina*, var. *variegata*. It is remarkable that all four examples of variegation belong to
the same genus, *Pteris* :* but, probably, other genera
will before long come into the sportive category.

* See a paper by "Delta," in the *Gardeners' Weekly Magazine*,
Feb. 4, 1861.

Dartmoor and the Dart.

DARTMOOR AND THE DART.

LOGAN STONE, NEAR CHAGFORD.

NEAR the middle of July, having a little boy just recovering from hooping-cough, for whom change of air was recommended, I determined to take him to

Dartmoor.   There was much in the historic associations of that seat of ancient civilization, that region where Britons mined and smelted tin and copper, and where Syrian and Sidonian merchants met, and chaffered, and displayed the precious things of Asia, ages before Rome was born—much in this, I say, that had a strange fascination for me; and, besides this, it was a peculiar tract of country, where I expected to meet with unwonted forms of life, interesting to both of us. Accordingly my boy took his butterfly-net and insect-boxes, and I a vasculum for ferns.

It was in cheery mood, then, that we set out from Marychurch on that brilliant July morning.   The long trailing twigs of the wild clematis were spreading in profuse and elegant freedom over the hedges and rocky slopes of the Teignmouth road, and hanging nearly to the horse's feet; and the dark and glossy leaves of the beeches almost intermingled as the long pendulous and patulous branches met overhead, as we turned off towards Hele.   The fetid iris, a plant so rare generally, so common here, showed its pale violet blossoms, in contrast with the crimson stars of the rose-campion; and all along the lower level of. the banks shone clusters of the sweet germander speedwell, always charming in its bright azure hue, and here, most poetically, named by the peasantry, "angels' eyes."   Groups of tall foxgloves rose from out of the rank herbage; and high up in the hedges there was a constant succession of dog-roses and honeysuckles, adorned with numbers of the

elegantly-banded hedge snails, on the scout for their
dewy breakfast, and looking like fantastically-painted
fruits clinging to the twigs.

Through Newton Abbot we roll, and remark with
surprise how fast it is increasing on every side, on
towards Ashburton, the scenery becoming richer and
more varied. It is market-day at Newton, and
numbers of farmers' daughters and wives, comely
Devonian dames, are riding to market with cloth-
covered panniers attached to their side-saddles, well
filled, doubtless, with butter and eggs, poultry-ware,
or whatever mysteries besides Dame Dobbins is wont
to carry on such occasions. The hedgerows are still
gay with flowers; the abundant yellow vetchling, two
species of St John's wort, the toadflax, precursor of
autumn, and hawkweeds, supply the golden colours;
the foxglove and the campions the crimsons; and the
blues and purples are afforded by the profusely elegant
racemes of the tufted vetch, resembling pendent
clusters of grapes in miniature. We miss the polystichum, our most common hedge-fern, and instead
of it the filix-mas spreads its great crowns along the
road-margin, and the tall brake crowds up among the
tree-trunks of yonder wood.

Peeps of the Moor, at least its south-west corner,
break upon us at various turns of the road, blue and
shadowy, like a huge rampart, ever becoming, as we
advance, more and more palpable. At length we
enter Ashburton, between lofty walls, gray, weather-
worn, and lichened. Old, old walls are these; for the

crevices of the courses are thickly studded with dense
tufts of the familiar wall spleenwort fern;* so thickly,
that viewed at a slight angle the whole surface of the
upper wall seems covered with the feathery fringe.
The plants have the appearance of great age; they
are very luxuriant, but this is the result of their long
growth, for when examined they have a very different
character from the long and broad-leafed succulent
crowns that adorn the hedges at Daccombe. These
are narrow, thin, dry and wiry, and far more plumose,
and of a golden-brown tint of green. The great snap-
dragon also throws out its crimson spikes from the
walls.

Here, then, in this quiet old-fashioned place, one of
the four ancient Stannary towns, and boasting the
honour of having given birth to Gifford, the quondam
editor of the *Quarterly*, we took up our head-quarters.

One of our earliest walks was to the romantic Dart
at Holne Bridge. Leaving the town by the opposite
end of the long street to that by which we had entered
it, we found the walls again clad with the *trichomanes*,
if possible even more luxuriant, and accompanied with
other ferns, as the wall-rue, the common polypody,
the ceterach, the maidenhair spleenwort, and the
hart's-tongue,—all, except the last two, particularly
fine. Indeed, we cannot walk a mile without per-
ceiving how pre-eminently this is a region of ferns.

It was a pleasant excursion, that early morning
walk, with the brilliant sun drinking up the dew-

* *Asplenium trichomanes.*

drops, and the lark singing overhead, recalling Shak-
speare's beautiful fancy :—

> "Hark ! hark ! the lark at heaven's gate sings,
> And Phœbus 'gins arise,
> His steeds to water at those springs
> On chaliced flowers that lies;
> And winking Marybuds begins
> To ope their golden eyes."

As soon as we are well clear of the houses, as if
the curtain of a stage had risen, we open in an instant a magnificent panorama. The country to the
south and west stretches up in lovely slopes, mapped
out in parti-coloured fields to the sky-bounded summits of the hills, divided by hedgerows and hedgerow
elms, with cottages nestling in the bottom, whence
the brawling of the little Yeo comes to the ear, as it
hastens on to join the Dart. We trace the receding
valley on towards Buckfast Leigh ; and far beyond in
the same direction Brent Beacon, a fine purple cone,
towers above the general level. The voices of hay-
makers taking advantage of the promising day, and
the sweet odour of the turned hay, tell what is going
on behind these impervious hedges of hazel and beech.
Before us the dim gray tableland of the moor, or at
least the south-west tail of it, lies under the softening
haze of the morning.

On the right the rugged rock shuts out the prospect,
for the road has been scarped out of the living shale,
but the ever growing shrubs and their gnarled roots,
with creeping and trailing plants, conceal the rough

stone with a graceful drapery. Half hid by a pendent curtain of ivy, we noticed a shallow cavity—I can scarcely call it a cavern—in which was growing a clump of navelwort,* of more than usual luxuriance even for these Devonshire lanes. The leaves, stalk, and flowers were all most succulent, not a blossom faded. One of these flower stalks, out of a clustered group, I measured; it was half-an-inch in diameter, and exactly thirty inches in length from root to point.

A little further on, from the same rocky barrier, a spring gushes out breast high, and is received into a stone coffer, like the new drinking-fountains of London streets, the water as clear and cool as a thirsty pedestrian could desire; nor was our relish of it diminished by the fact that we found in its clear recesses other revellers,† thorough teetotallers all.

Presently a rushing brook passes beneath the road, and, being collected into a wooden channel, crossed with great flags for gangways, hurries away to a clacking mill. It was pretty to mark the dense subaqueous forest of water-blinks, how its long tufts waved and quivered in the black rapid stream, beneath the shadow of the tall meadow-sweet, most delightfully fragrant, that profusely bordered the water. This, however, was a deviation through inadvertence; recovering the road by a cross lane, we found the columbine

* *Cotyledon umbilicus.*
† *Gammarus fluviatilis, Ancylus fluviatilis, Limnæus pereger, Helix rufescens.*

growing luxuriantly, said to be most elegant with its
tall blue blossoms, but now out of flower; and the
tutsan, the finest and rarest of our truly native St
John's worts. The beautiful spotted bastard-balm,*
rarer still, was also here in full bloom, spangling the
hedge with its large white flowers, blotched with
purple on the lip. This is probably the identical spot
in which it was found by Sir James Smith, who gives
as its locality "a mile from Ashburton, on the road
towards Plymouth."—(*Engl. Bot.*)

Now we are in the valley of the Dart; and those
hills that slope upward on our left, so densely covered
with dark timber, making a noble appearance, are the
Hembury woods, on the other side of the river. A
little farther, and the whispering sound of running
water is heard, which momentarily increases in fulness.
We know it is the Dart; but the bank of the road is
high, and curbs the prospect. Here is a gap in the
hedge; we look over into a dark glen, shadowed by
tall trees, and crowded with foxglove; yonder is a
bottom of deeper greenwood obscurity, whence the
denser beech woods rise on the opposite hill. There
is the seat of the sounding stream, but we see it not.
A few yards more, and we climb over into the solemn
grove: the whisper has become a roar, and there
sweeps the lovely river. Just where we have struck
it, a ridge of broken rock crosses its bed; above this,
there is a glassy expanse, mirroring the trunks of the
bordering trees and the black shadow; below, a broken

* *Melittis grandiflora.*

2 B

foamy torrent : there is a short but pretty plunge over
the dividing ridge, with a deafening noise, that combines the whisper with the roar. The views upward
and downward are both beautiful ; the latter especially, where the tall overspreading trees—*patulæ fagi*
—shadow the whitening current in a long receding
vista.

I was struck with the resemblance borne by the
bank of the stream to the sea shore ; the shale crops
out, and this is worn by the rushing flood into smoothed,
but irregularly angular, projections and hollows, exactly like tide-pools, while the sand lying here and
there, washed into ripples, increases the *vraisemblance.*
Nor was the illusion altogether dispelled by the vegetation ; for the *Osmunda regalis*—the royal flowering-
fern—which grows all along the very water's edge,
presented a strange likeness to a sun-dried sea-weed,
for the clumps growing on the points of the shale, and
out of the sand, had formed great compact masses of
black wiry roots, from which in many cases almost
all the soil had been washed away, and which threw
up only a few stunted and deformed fronds, in no
adequate proportion to the size of the root-masses.

A tiny brook comes down the left bank, and pours
its little cascade into the stream ; and just here we
find some fine tufts of the hard-fern,* growing out of
the sand. It is a fern which we do not know around
Torquay, but we became subsequently familiar enough
with its long pinnate fronds. The azure heads of the

---

* *Blechnum spicant.*

sheep's-bit scabious, an old acquaintance, were peeping everywhere.

Pursuing the high road a few rods farther, we arrived at Holne Bridge, a solid granite structure that spans the Dart, well clad with ancient ivy and ferns. From the roof of its main arch depend stalactites, nine inches or more in length, formed by the free lime of its cement, each one bearing evidence, in its terminal drop of clear lime water, surrounded by a thin and brittle coat, of present increase.

Here we loitered for hours, lounging under the cool shadow of the bridge, admiring the *Jungermanniæ* and mosses that so luxuriantly cushioned the beetling hollows of the rock;* or gazing alternately up and down the river, from above the bridge and from below —up, where the stream comes brawling and sparkling down, curving round from the heights of Buckland woods, a mass of white foam, broken by dark rocky boulders; below, dark and mirror-like between tall beech woods on either bank; or peering, with that strange fascination which impels one with an ever increasing power to plunge in, on the deep hollows, so black and yet so smooth—so treacherously and invitingly smooth—" like sin," as my little son said.

I did not estimate the height of the bridge; but from recollection I should think the arch about twenty feet above the water as I saw it. Yet I was assured that occasionally the river is so swollen that the flood completely fills the arch. It must be a grand sight

---

* *Splachnum ampullaceum* was particularly large and fine here.

to see such a body of water, far exceeding the limits
of the rocky bed, and filling a wide expanse of the
bordering valley.   It would then be a large river, and
worthy of the lovely scenery by which it is encom-
passed; and, pouring down with its characteristic impetuosity, an impetuosity justifying its name of "Dart,"
—it must, I say, be a grand sight. The floods of this
river are extraordinarily sudden. After heavy rains
in its elevated sources, far up in Dartmoor, of which
perhaps not a drop has fallen here, the swollen waters
come bearing down, all without warning, in the form
of a mighty wall, like the bore of the Ganges in the
monsoon.

Like the Lea, near London, the Dart has the reputation of being peculiarly fatal, not a year passing
without one or more persons being engulphed by it.
This belief is embodied in a provincial distich :—

> "River of Dart ! River of Dart !
> Every year thou claimest a heart."

The swiftness of its course, and the great unevenness
of its bottom, full of treacherous holes and pits, combine with these sudden floods to give it such a character.   People speak of its shrill wailing rushing
sound as "the river's *cry;*" and say that during the
stillness of night they hear it a long way from its
banks.

Some little distance above the bridge the shale
forms a noble precipice, rising with a rough face almost perpendicularly from the water's edge.   It is
crowned with fine timber, which descends on either

side to the river. A romantic story cleaves to this
cliff. A too-confiding maiden, deeply wronged by her
lover, leaped, Sappho-like, from the summit; and
hence the sheer descent bears the name of the Lover's
Leap.

We heard a good deal of Widdicombe—"Widdicombe in the cold country;" "Widdicombe in the
Moor;" "Widdicombe in the Dartmoors;"—"you
must see Widdicombe by all means." Our curiosity
was excited, and we set out. We wished, too, to see
Widdicombe Tower, that tall embattled structure
noted for its architectural beauty, and not less for its
having been the scene of a most fatal elemental war.
A terrible thunderstorm burst into the church during
afternoon service, Oct. 21, 1638, and besides crushing
and defacing the material building, killed four persons, and more or less wounded sixty-two. One of
these was slain with unusual circumstances of horror.
"Another man," says the local record, "had his head
cloven, his skull wrent into three pieces, and his brains
thrown upon the ground whole; but the hair of his
head, through the violence of the blow, stuck fast to a
pillar near him, where it remained a woeful spectacle
a long time after." Tradition affirms that the sufferer
was sacrilegiously playing at cards in his pew at the
time; and that it was the Prince of Darkness himself
who, riding on the thunderbolt, accomplished the
vengeance.

Well, Widdicombe and its tower must be seen.
So we set out in the hard gray morning, leaving Ash-

burton by another outlet, a street sufficiently sordid, where the little river Yeo purls alongside the causeway, and disappears from the eyes of passengers to wash the back-doors of the old quaint houses, and where dingy women are ever going up and down steps with pitchers and kettles. We get beyond houses and bridge, and pass as usual between walls bristling with feathery ferns, evidently the invariable in the outskirts of this little town.

The ground now rises fast, leading us through a high banked road or lane (I scarcely know which), where the male fern is profuse and fine, and the brake is also large and succulent, and delicately green. Ferns appear to have their proper domains. Presently these disappear, and it is all the black spleenwort with enormous fronds, so elegant in their triangular outline, and dark glossy greenness; and by and by nothing but the hard fern springing out of the loose stone walls. Oaks stretch their arms above the road, and make an almost continuous canopy, through which the twinkling sunbeams struggle and play, with soft pellucid yellow-green light; anon these give place to sombre firs, dull, stiff, and skeleton-like, the plantations of Welster.

A conspicuous object all the way up has been the works and tall scaffolding of a mine on a hill to the right. We are told its name is Druids' Mine, and we at once think of the Phœnician times; but no! it is a thing of yesterday; a London speculation of some five years' starting.

We toil up past the mine, and, getting behind and above it, have a fine view of the valley to the east, and of the country toward Newton. A tall hedge of dwarf beech bounds the road on one side, and furze thickets on the other. Here my entomological son got his first prize; for we saw specimens of the beautiful little Green Hair-streak, which presently became sufficiently abundant, and continued the characteristic butterfly up to the moor.

The driver pulls up, and, pointing to the left, says, "This is Welster Beacon." We look, and see nothing to tempt us to get out. "No; but just over the brow you'll see Buckland Beacon, about a quarter of a mile further on; and that is one of the old Tors, where you'll see a good sight." We dismount, clamber over a gate, and are on a wide rounded hill, covered with button-like clumps of mingled heath and furze; a broad grassy path leads us to a rough stone wall, and there the Tor stands revealed in its majesty, with its conical crown of gray stone.

We are soon on the summit, gazing with eager eyes on the almost boundless expanse which stretches away on every side. It is a noble prospect. Unfortunately there is a haze in the air, which diminishes, though it does not destroy, the distinctness of the several features. The eye is first attracted by the winding, sparkling. Dart, which seems to be beneath our feet, as we trace it coming in from the distant moor, and then encircling in its twining embrace the lovely Holne Chace, which, with its rich timber-woods and

its curving drive, looks from this height just like a
villa garden with its gravel walk. Thence the charming river is seen, white with foam (I unconsciously
listened for its babblings), meandering away to the
south, till the tall woods conceal it. There too was
Buckland Mansion, dwindled to a toy-house, and the
gray church tower rising from the embosoming trees
a little behind it. Beyond and around were the map-
like fields, of many hues, chequered out with hedges,
and far on the south-west horizon the blue Brent Tor.

In the foreground to the south there is Answell or
A's'ell Rock, a very remarkable Tor, crowning an
eminence considerably inferior in elevation to that on
which we stand. Viewed from hence, it takes the
form of a colossal lion couchant, with a second granite
mass, that looks like an armadillo, or, if you please,
the great fossil tortoise of the Siwalik, marching up
behind him. The buildings and tall shaft-works of
Druids' Mine, a little to the left, stand out bold and
prominent. All the east is dim, and obscurely empty;
but on the north rises Hay Tor, a fine purple cone of
lofty stature, and the wide gray moor stretching away
to the westward, a dreary, tabular, unvarying mass.
The sides and slopes around our feet were spotted
with a few semi-wild sheep, which we fondly con-
cluded made the well-flavoured Dartmoor mutton,
that the reader doubtless wots of; but the driver
subsequently undeceived us as to this point. The
true breed is found only on the moor proper, which
we have not yet come to.

We descend perilously, for the granite is angular
and rugged, and the short wiry turf particularly slippery. Besides this there is little vegetation, save the
tiny bedstraw, profuse in white bloom, crowding the
rocky crevices, and the rosy white stars of the English
stone-crop, and a twine or two of ivy mantling the
gray stone, and abundant lichens completely covering
the old weather-worn surfaces. A pretty little yellow
cinque-foil studs the short turf with its flowers, and
on the rounded hill there is the furze. This is singularly close and dense, forming rounded knobs, as
even and shapely as the well-shorn shrubs in a Dutch
garden; the wind doubtless having performed the
shearing operation here.

The mass of granite has a very noble and impres-
sive aspect, as we gaze up upon it from the point
where it breaks out of the turf. It takes somewhat
of the form of an immense table, rising to a low cone
in the middle, or something like a huge loaf that has
sunk in the baking. The vast block of primeval
stone, naked and hoary, is imposing in its severe
simplicity, as it stands out broad against the sky, with
no object around on the bare hill to distract the
attention; and it derives a great extrinsic interest
from its associations with early, perhaps even pre-
historic, times. This is Buckland *Beacon;* and I
picture to myself the pile of combustibles, which in
seasons of danger or other expectation was wont to
stand on that massive table, and the wild form of the
appointed watcher, possibly the bearded Druid him-

self, rushing up the hill with his flaming brand, and
hurriedly lighting the pile.   Then I see the fire lick
up the heap of crisp and sun-dried leaves, catch the
faggots of furze, and, overpowering the cloud of blinding smoke, burst forth into a grand pyramid of flame,
whose leaping and curling tongues rustle the scorched
air, and roar like a tempest, while the red glare falls
on the lone waste around.   Then in an instant the
attendant, gazing wistfully into the darkness, sees the
answering fire blaze out on the summit of Rippon
Tor, and presently, it may be, another spark, just discernible, tells that a more distant beacon has caught
the signal, and sent it on ; and he knows that before
he can descend the slope and reach his lone hut, the
whole of the wide moor will be studded with points
of flame, and all the dwellers in the wilderness will
have received the communicated intelligence.

Perhaps the ordinary occasion on which these
signal fires were lighted may have been the approach
of the Tyrian galley coming to trade for tin.   The
estuary of the Tamar has been shown on almost irresistible evidence to be the ancient Iktis, the port
whence the Phœnician tin trade was carried on.
Thither the accumulations ot " silver, iron, tin, and
lead," (Ezek. xxvii. 12,) which the mines of this rich
metalliferous region* were yielding to the labour of a
not unskilled and uneducated population, were brought
for sale and barter; and thence the purple and fine
linen of Tyre, the elegant manufactures of the Sido-

------

* See Appendix; *infra.*

nians, the gold and precious stones and spices of India, and perhaps even the silks of distant China, were distributed through Britain.*

The arrival of the periodical galley from Tyre would then be an event which would put in motion the entire mining country; and we may well imagine that a series of signals would be concerted, by which it should instantly be known. These lofty Tors were admirably calculated for such a telegraphic system. Yonder blue mass in the south, Brent Beacon, overlooks Plymouth Sound and its approaches, and the appearance of a sail on the distant seaward horizon would not fail to be noted in fine weather. "What would be done," do you ask, "if the galley made the coast in misty or hazy weather?" I suppose in such case a signal fire would be lighted on the heights around the Sound, as soon as it became clear.

The flame from Brent would be responded to by Buckland in this direction, and by Sheep Tor in the west; from the former the news would flash to Rippon, or Hay Tor, and be known then over the whole eastern region; from the latter it would be caught by Tor Royal, and thence by Crockern Tor, that seat of the Druidical legislature, and by Great Mis Tor, whence it would dart to Fur Tor and Yeo Tor, and to Watern Tor, that overlooks that wild lonely lake whence the Devonian rivers take their rise, and to Cawsand Beacon, from whose towering top the flame would be visible at once from both the northern and

* See Heeren's Historical Researches, *passim*.

the southern sea. The distinguishing title of " Beacon," which immeasurable tradition has connected with some of these eminences, in nowise differing from their sister Tors, indicates the use which they subserved.

Reluctantly we leave these imaginings, and turn our backs upon the ancient pile, Willie first rooting out a crown of the hay-scented fern from the foot of the " clatter," as a mass of granite so situated is provincially termed, and bagging a tiny lamellicorn beetle, and one or two more of the hairstreaks, that were fluttering over the furze-clumps.

We ride on to make up for time lost in dreaming dreams of three thousand years ago, between the old roadside walls, made of granite blocks, rounded now by the effects of frost, and wind, and rain, and completely encased in lichen coats of many hues. Here the broad buckler-fern becomes the predominant form, remarkable for the very peculiar aspect produced by the recurvation of the edges of its pinnules, as if the plant had been just blighted by some baneful application, while yet retaining its succulency and greenness. The hard-fern and the male-fern, however, still occur also, the former unusually fine and well grown. Great agarics, too, and agaric-shaped *boleti*, are conspicuouly numerous all along the road in this region. We pass through a gate, and are told that we are on the Moor.

A wide, desolate, heathy waste it is, undivided by fences (at least where we are travelling), but intersected by the high road. It did not oppress me with its loneliness; for the sun was bright, and horses were grazing, and sheep bleating, and butterflies were fluttering, and large handsome dragon-flies were hawking to and fro on rustling wings over the rain-pools in the sand-pits.

So we jog along, opening and closing in turn one and another of those wild fantastic heights, whose outlines ever change as we pass, and admiring the playing lights and shadows over the dark purple surface of the moor, as the clouds borne by the breeze expose or conceal the sun. The brooks enliven the waste, pouring along in haste, and sparkling among the granite boulders of their beds. Several of these cross the high road, the feeders of the Webbern, itself a tributary of the Dart, affording clear cold draughts to our thirsty throats, but with a somewhat ominously brown tinge and moory taste.

At length we emerge on a wide and most lovely valley; farms and hamlets scattered about its slopes, each surrounded by its half-concealing groves of dark trees, look like oases in the wilderness. Some distance up the vale, which bends away from sight between the hills, we see the lofty church-tower of Widdicombe, a fine gray square pile, of what is called the Perpendicular style, with four pinnacles at the corners. It rears its head far above the village, which seems to nestle among the embosoming groves, as if desirous to be unseen. It is the frontier of civilization; far beyond all is the wild and naked moor. Imagination rests awhile on those sweet slopes,

mapped out into fields, and spotted with clumps of
noble trees, most of them ancient sycamores; and
then, gathering its garments tightly round, strides
onward into the unknown waste that lies among those
sweeping moory mountains.

We did not go down into the village, but having
satiated our eyes and our fancies, more ignoble, but
not less clamorous propensities put in their claims
for attention. Sitting down beside one of the gurgling
moor-streams, we spread our frugal provisions on the
short turf, pic-nic fashion, and ate and drank with
gusto. Then strolling down the rivulet a few paces,
I found it dilated into a shaking bog, where footing,
from clump to clump of furze, and from tussock to
tussock of grass, was both scarce and precarious. I
had been told that a singular and highly interesting
plant, the sun-dew, that has the power of capturing
unfortunate flies which rest on its leaves, grows in
these elevated bogs, together with some other scarce
plants, as the asphodel, and the beautiful little bog
pimpernel. I found, however, neither of these, but
the bright-eyed laughing forget-me-not, whose appeal
I chose to read as made on behalf of the charming
Vale of Widdicombe in the Dartmoors; and this I
do not think I *shall* very readily forget.

Just as we were setting out to return, we had a
curious evidence of the lonely and habitually undisturbed character of the place in the conduct of a little
bird. A skylark descending came down close to us,
singing as it came; and lowering itself on wide-spread

wings, slowly reached the turf within a yard or two
of our feet, without the slightest manifestation of
fear, or even consciousness of our presence. A more
unwelcome creature, but by reputation highly characteristic of the moor, crossed our path in the form of
a viper, which the driver speedily demolished with
his whip. It was the red variety, very distinctly and
elegantly marked. The common people here call
these reptiles by the strange name of *cripples;* a most
unaccountable appellation surely. Musing on its
etymology, it occurred to me that possibly it may
look back to the Roman dominion, and may be the
Latin *coluber*, the letters *l* and *r* being anagramma-
tically interchanged.

Another excursion of much interest was one that led
us, by Sharpitor and Dartmeet, to Wistman's Wood.

After striking the Dart at Holne Bridge, we cross,
and proceed through Holne Chace, a picturesque wood,
from which the heavy timber has been weeded out.
Between the road and the river our attention was
directed to the enormous stumps of trees that had
been felled even with the ground. The driver—the
usual authority for tourists—informed us that the
proprietor a few years ago cut timber from this property to the value of £60,000, which he bestowed as
a fortune upon his daughter. The magnificent mass
we are looking at was the base of the " queen" tree;
the " king," of still superior vastness, is on the opposite side of the river.

Another quaint little bridge crosses the winding Dart at a most charmingly wild spot, having passed which, we rapidly ascend, and approach Leigh Tor, (pronounced Léftor,) a remarkable ridge of granite, crowning an eminence, which runs for some distance from east to west. We alight to examine it. It consists of enormous blocks, so thrown up as to present a sharp angular keel to the sky, of course irregular and broken, and much more vast than it looks from the road. There was something very noble in its rude magnificence, as we stood against it. The common polypody in profuse abundance was creeping about the sides and points, making thick carpet-like draperies of its many bright green pinnate leaves, while in the dark crevices and obscure holes, overhung by the leaning blocks of stone, the brood bucklerfern was fine, and wood-sorrel displayed its threefold foliage.

The woods of the Chace, through which we are still winding, are very beautiful; they are chiefly oak, but there is a great deal of birch, sufficient to give quite a preponderating character; and the light mobile foliage of this tree, " the lady of the woods," contrasts well with the massiveness of its lordly companion. Ferns, too, are very abundant and luxuriant, especially the *blechnum*, which I never before saw so fine.

The travelling becomes increasingly steep; the prospect behind and on either side wider and grander; the country through which we are passing is more

than usually picturesque ; old fashioned farm-houses,
a primitive inn, an antediluvian smithy, and people as
unsophisticated as the region ; little brilliant streams
crossing the road at every turn ; fence-walls of huge,
unformed granite blocks ;—these at length bring us
slowly and toilingly upon the moor.

And now we are in the midst of the Tors. Here
they are on every side, Pelion upon Ossa, Ossa upon
Olympus ! Away on the left, or, as a seaman would
say, broad upon the larboard bow, rises from deep
valleys a vast mass, of great regularity in outline,
terminating in a sharp cone of gray rock. Other
hills are beyond and on each side, but there is a
grandeur in this which arrests the attention. Presently the road turns, and we see that we shall have
to skirt the very flank of the mountain. A passing
peasant, of whom we inquire its name, tells that it is
Sharp Tor, or, as he pronounces it, Sharpitor. As we
pass, we perceive that its broad side bears antique
remains, and jump out to explore. Yes; here are
indubitable relics of man's handiwork. We are
standing in the midst of a circle, about fifteen yards
in diameter, as I measured by pacing, composed of
great primary rounded blocks of granite. Besides
these, however, there are the débris of the elevation, in heaps of smaller stones which lie at the
base of the primaries, making the walls in places
two yards or more in thickness. The primary stones
are placed in contact, not separated by interspaces,
and therefore this was a dwelling-house, not a

temple or sacred enclosure. There is an avenue of parallel stones leading from the circle away towards the south-east, that is, to the nearest point of the Dart.

The great size of this circle, as compared with the ordinary hut-circles, which rarely exceed nine yards in diameter, is remarkable. It is not sufficiently wide to enclose a village, nor to serve as a " pound " for cattle. It would rather seem to indicate the residence of a chief, but I did not notice any second circle within the enclosure.*

The bright, laughing sunshine and the dancing butterflies were not favourable to those solemn reveries which the same scene under the silent moonlight, or in a twilight storm, would have conjured up. Yet it was not without a feeling of awe that I reflected on the hoary ages that had passed away, since a man of like passions with me brought home his loved bride to that now lonely dwelling, and " his young barbarians all at play" made the air ring with merriment as they ran around those gray wall-stones, or chased the identical forefathers of those very butterflies over the moor. " How long was that ago ?" How long ? Who knows ? Perhaps when Jephthah

---

* I took for granted that I should find full details of this remarkable circle in Rowe's " Perambulations," which I referred to as soon as I arrived at home, but to my surprise he does not notice any remains on this side of the hill. Had I anticipated this silence, I should have made a more minute examination ; the notes I have recorded above were, however, jotted down in my note-book upon the spot, and may therefore be depended on.

was offering up his burnt-offering; perhaps when
Caleb was carrying the cluster of Eshcol.

Yar Tor rises just opposite to Sharp Tor, on the
right hand of our road. This eminence is crowned
by a strong wall, forming a square enclosure, formed
partly by nature and partly by art. On the north-
east slope we saw "the remains of some hut-circles,
and the ruins of a kist-vaen," which are there; but
time pressed, tokens of a storm lurked in the horizon,
and I specially deprecated our being disappointed of
seeing Wistman's Wood. We were therefore con-
strained to hasten past these interesting antiquities
without examination.

From the skirts of Yar Tor, looking south, the
view is suddenly opened by our having left Sharp
Tor in the rear; and the rounded hills which form
the valley of the Dart take majestic forms, and have
an aspect of sullen grandeur, but the river itself is
not here visible.

Soon, however, we see it before us; and descending
abruptly we come to a small bridge, by which we
cross the East Dart, just before its junction with the
West Dart; the spot itself and the bridge bearing
the name of Dartmeet. The view up the east branch
is extremely fine; the stream being very swift, and
full of rocks, the wild riotous turbulence, the dashing,
flashing water half lost in foam, and the high-toned
metallic cry, take strong hold of the imagination.
Looking up the west branch, and down the united
stream, the aspect has the same character.

Beside the road which rises from the bridge, the elegantly cut lady-fern suddenly becomes so abundant as to be characteristic, throwing out its arching fronds from the walls, mingled, however, with the everywhere profuse hard-fern and brake.

Far up on the naked moor, a little black object appears, to which our attention is drawn. We might have imagined it a traveller's portmanteau dropped from a carriage, or a bit of india-rubber left by a sketcher, so minute it looks on the desolate mountain side, with no other object near with which to compare it. We are told it is a school; a school endowed by two benevolent ladies. A school, forsooth! we should almost as soon have expected a coffee-house or a jeweller's shop. Where do the children come from? we asked. At last we reach it, the road leading by the door. It is built of corrugated iron, like a temporary railway station, and however we might have doubted the call for such a building, there are the urchins within. Through the open door we see them sitting at the desks, gazing furtively at us, doubtless envying us freedom and a ride. The school rejoices in the euphonious appellation of Brimps.

Not far from this, our civil and intelligent driver took fire at the occurrence of a wall, which he had not seen before, not having been so far from home for a twelvemonth or more. It is a noble wall, well-built of massive blocks of granite split out of these endless Tors, and put together without the least bit of cement, the solidity and weight of the pieces

affording the requisite stability. Wide interstices
between the blocks are left purposely, that the force
of the wind may not act unbroken. What a shrill
wail these varying orifices must utter in a high
autumnal gale! The immense size of the blocks,
contrasting with those of the ancient walls around,
shews the resources of modern art; for this wall is
built by Government. It was this that stirred the
wrath of our friend, who protested eloquently against
the encroaching policy of the Duchy of Cornwall,
robbing, as he averred, the poor moormen of their
scanty pasture. At first we could not see the object
of the enclosure, but presently a healthy plantation
of young firs within the grim fence explained it.

All before us spreads the wide moor, ever wilder
and wilder. Solitary farms smile at distant intervals
in the bottoms. One of these, Sherbiton, prettily
surrounded with trees, by a brook, specially attracted
me. Abundant springs of water occur everywhere;
the moor is evidently a Jamaica, a "land of springs."
Far away in the west, we see Princeton and the
War Prison, which at one time held ten thousand
prisoners. Grim and gaunt, it stands on the naked
mountain-side, encircled by its gray granite walls.
On the right of this uninviting object, but much more
distant, a magnificent Tor rises against the sky, striking the eye by its noble outline and manifest elevation, pre-eminent above its fellows, and exciting
curiosity to know its name. At length a peasant
informs us that it is Great Mis Tor.

This is one of the loftiest peaks on the moor, and
appears to have been a chief seat of the Druidical
astrolatry, at least if that etymology of its name be
sound which derives it from Misor, the British moon-
goddess, answering to the Ashtoreth of the Sidonians.
The out-croppings of the granite on the summit of
the Tor are described as taking magnificently massive
forms; and, on the top of the loftiest, at an elevation
of 1760 feet above the sea-level, there is one of those
singular circular pans which have provoked so much
discussion as to their origin. From their situation,
and the regularity of their form, it has by some anti-
quaries been maintained that they were connected
with sacerdotal service, used either for the washing
of the sacrifices or the ceremonial lustrations of the
priests, and that they were excavated by human skill
and labour. Others believe that, beautifully regular
as is their shape, they are merely the result of natural
decomposition. If so, however, they are not the less
likely to have been pressed into the service of the
dominant idolatry. The one on Mis Tor, commonly
known as Mis Tor Pan, is remarkably smooth, and
perfect in its outline, about eight inches in depth,
and a yard in diameter.*

However, this was far beyond our present limits.
We jog on by Hockaby Tor, a crowned eminence on
our right, of no great elevation, but interesting because on its slope there are several hut-circles distinctly visible from the road, distant from it, indeed,

* See the account of the Punchbowl, on Lundy; p. 97, *supra.*

only a few rods. We ford Oxlake Brook, a little feeder of the West Dart, which latter we see winding among the hills at our left, running nearly parallel with the road, and forming brilliant little pools here and there under the midday sun. The slopes from Sherbiton are covered with boulders, which *may be* ruins, but which appear, from their wild confusion and ruggedness, to lie where they were protruded in some primeval convulsion of nature. Their number imparts an unusual aspect of savage desolation to the scene. Here, on a granite block, was sitting a yellow-hammer preening its bright plumage in the sun, and all unconscious of sadness. The local name for this familiar little bird is "gold-gladdie;" a pretty appellation, well suited to the appearance of a gay-coloured bird in a lonely and desolate region.

Our road now is bounded on one side by the wall of Dinnabridge Pound. Within the enclosure near the gate there is a lichened chair of granite, evidently of great antiquity, which tradition asserts to have been brought hither from Crockern Tor. The Pound itself is comparatively modern; it is a large enclosure, surrounded by one of the dry-stone walls so universal here. It owes its origin to the customs of the moor. The entire forest is farmed out under the Duchy of Cornwall to certain persons termed moormen, who receive cattle for pasture, affixing to each animal "the moorman's mark." On one day in the season, of which he gives no notice, the moorman drives all the cattle on his quarter to a given spot, and impounds every one that has not his mark.

The sheep and cattle, though allowed to wander apparently without restraint, are said always to keep company with those of their own flock or herd. We could see this to be true with respect to the sheep; though we passed a great number in our course over the moor, we never saw those marked with one colour or set of letters mingling with those otherwise marked; they associated with their own exclusively. These are true " Dartmoor muttons," a little breed, which you would think to be lambs. The sun was powerful; and the poor things in this shadowless region were fain to creep up under the north side of the stones, and even to the sides of the shallow pits whence turf had been dug, as an apology for shade, though even when they lay down more than half of their bodies were in the full blaze.

Just beyond Dinnabridge Pound we see a fine Tor, which bears the name of Belevor, one of the numerous eminences in which the name of Bel or Baal is traced; a remnant of the Oriental idolatry which spread its baleful wings over this region in primitive ages. The country hereabout is full of aboriginal monuments, and the Tors rise thickly on the right and in front of us; the mountains bristle with rocky peaks and points, like monumental stones in a churchyard; grim memorials of the long, long past. Conspicuous among them is Longaford Tor, with its sharp conical crown, and in the foreground Crockern Tor, once a place of renown.

On this bleak and desolate mountain, in the midst of these hoary stones, with only the sky for a canopy, were it clear or cloudy, bright sun or driving rain, soft zephyr or howling storm, met the ancient Stannary Parliament. Each of the four Stannary towns, Chagford, Ashburton, Plympton, and Tavistock, sent twenty-four burgesses to this assembly, when summoned by the Lord Warden of the Stannaries. They enacted laws, which, when ratified by the Lord Warden, were in full force in all matters between the tinners, " lyfe and lym excepted." Probability and tradition, however, assign a far higher sanctity to this spot, as the seat of Druid legislature. The Stannary Court was but a form, comparatively modern, of an assembly which had been wont to meet there from earliest times. Polwhele says that in his time there were the president's chair, seats for the jurors, a high corner-stone for the crier of the court, and a table; all rudely hewn out of the granite of the Tor, together with a cavern, which may have served as a dungeon for the condemned. A moorland patriarch who had known the spot for threescore years, told Mr Rowe, in 1835, that he perfectly remembered the stone chair, which was ascended by four or five steps, and that overhead it was protected by a large thin slab of stone. There seems to be little doubt that this chair now stands in Dinnabridge Pound.

A strange feeling pervades the mind as one stands thus amid the memorials of a hoary and dim antiquity, and calls up the gathering assembly, with its stoled priest and its sucking lamb, and perhaps other sacrifices, from which imagination hides her face, dripping blood on these huge stones, perhaps at the very time that Samuel was judging Israel in the circle of Gilgal, and obtesting at the stone of Eben-ezer. The nakedness and utter desolation of the region still and solemnise the mind, notwithstanding the cheerful sun and the swift chequering clouds. In the shadows of evening it must be an awful place, when fancy would people it with the shadowy ghosts of four thousand years, and every stone would take form and life.

Again we descend to the river at a place called Two Bridges. There is, indeed, but *one* bridge that I could see or hear of; and the only way I can account for the name is by supposing that there existed, coevally with this modern structure, one of those very ancient ones, usually described as " Cyclopean," made by upright monoliths for piers, and enormous slabs of granite made to rest on these horizontally, the stability of the whole depending on the immense weight of the blocks. Several of these interesting bridges yet remain to attest the solidity of their construction, particularly one over the East Dart, near Post Bridge.

Close to the river, at Two Bridges, a country inn of more than usual pretensions, surrounded by its offices, gives animation to the scene. The sign, once doubtless a fine example of high art, representing a robed and bearded personage, life-size, now, from the painting having never been varnished, and the colours having fled, presents an appearance ludicrously ghastly. An antique ostler, curiously in keeping with the associations of the place, appearing, we inquired our way to the wood, the object of our search, while the driver dived into the recesses of the bar, to get " summat." The man of horses, with a ready civility, and a free use of " yir 'annor," which suggested a Milesian education, pointed up the valley of the Dart to the slope full in sight, at what appeared to be only a few hundred yards distant, where a very circumscribed thicket of scrubby furze (so it looked) was growing. "An' that," said he, " is Wistman's Wood ; an' it 's a mile and a half's walkin'."

My little son and I tramped away up the valley, the scrub still in sight every step of the way, which ever grew wilder and wilder. Through a dreary farm-yard we post, and the world is behind us. Our course is parallel to the Dart, which purls and rustles below, under the shadow, for some half-mile or so from the bridge, of magnificent dark beeches. The utter absence of recognisable objects makes distances strangely deceptive. Still wilder and wilder grows the moor ; here and there we suddenly find our feet press a shaking bog. No sign of animal life, but a pretty little butterfly,* which is numerous. Yes ; in

* *Hipparchia pamphilus ;* a variety much paler than the ordinary condition.

that little pool in the rocky stream, a trout leaps into
the air; another, and another; though the dimensions
of the pool are such as you might stride across, and
its depth seemingly fathomable with your hand.

The furze begins to take form, and to look like
dwarf trees, and far above frowns the blue peak of
Row Tor, around which dark clouds are gathering.
Now the granite blocks become thicker and more
numerous, till we find ourselves crossing a wilderness
of boulders, where we can proceed only by scrambling
from one to another, there being literally no way
between, the narrow interstices being choked with
brake and moss, and the stone-crop. I never before
saw a place which gave me such an idea of utter desolation. At length we reach a single solitary oak, the
outpost of the wood, and after a little more difficult
and dangerous scrambling over the blocks, we enter
the weird forest itself.

I despair of conveying an idea of the strange scene
by words. The granite boulders continue as close as
before, with the stunted and most gnarled trees
springing out of the interstices. It is said that these
narrow passages go down, down, to an unknown depth;
and some have thought that we see only the summits
of the trees, the trunks, of ordinary height, being
rooted in the earth far below. I did not think of
probing the treacherous crevices, which are reputed
to swarm with adders; but there is that in the aspect
of the trees which at once confutes this hypothesis;
not to speak of the absurdity of supposing that the

granite boulders have gradually accumulated since
the existence of the trees.

That these oaks and rowans are enormously old I
do not doubt. Those which I saw might have trunks
a foot in diameter on the average, and their height is
from twelve to fifteen feet. The branches are wonderfully twisted and knotted; the heads are scanty,
flattened, and wide-spread; and both trunks and
boughs are so thickly encrusted with dense moss, that
ferns grow profusely upon them.* The foliage is not
unhealthy; and I observed numbers of those globular
galls, like a boy's marbles, which have lately attracted
notice in South Devon. I saw no young trees, no
suckers, no acorns. Individual immortality seemed
to be conferred on these remnants of bygone times,
but nothing that indicated transmission of life to
another generation.

I should much like to see a section of one of these
old trunks; to count the concentric circles, and thus
obtain a clue to their actual age. At the Norman
Conquest the wood is said to have presented the same
appearance as it now does; and I should think it by
no means an unreasonable conjecture that these identical trees have witnessed Druidical rites. The explanation of the name, "Wistman's Wood," as *wise-
man's* wood, has been ridiculed; but, considering the
ancient form of "*wist*," (from the Saxon pɪꞃᴅɲ,) *to
know*, it is highly suggestive: *q. d.*, the Wood, or

* *Polypodium vulgare*, and *Lastræa dilatata* and *recurva* profusely.

Sacred Grove, where the *Knowing-man* performed his incantations.

By this time the heavens were gathering blackness, and from several points of the horizon those dark, ragged clouds were rising and hanging in tattered shreds, that tell of heavy rain. Mutterings of thunder, too, were audible; and notwithstanding that a thunderstorm in such a place as Wistman's Wood would have greatly augmented its melodramatic interest, yet neither on my own nor my sick child's account did I exactly wish to brave its results. We, therefore, hastened back to Two Bridges, casting many a wistful glance on the strange scene we had left, the like of which we might probably never see again. It took a peculiar aspect under the glowering sky; and the distant peak of Row Tor above, lighted up by a momentary gleam of sunshine, came out wonderfully fine against the black storm-cloud.

Some points of interest occurred on the homeward route. Several hut-circles were seen close by the road-side at Haxary, where the West Dart gleamed beautifully in a romantic dell. Then we reached Compstone Tor, crowned with a fine assemblage of granite rocks of that peculiar form known as the Cheesewring; enormous slab-like masses of varying diameters, piled one on another horizontally: I say "piled," for such is the appearance; though doubtless the phenomenon is the result of elemental decomposition on the horizontally stratified granite. This arrangement has a very magnificent effect.

Benjay Tor, a lofty hill, which presents a precipitous face of broken stone to the bed of the turbulent Dart, was a striking object, the river forming two deep and dark pools beneath, which are called Hell Pool and Bell Pool. But, I think, as interesting as anything that I had seen, not excepting even the architectural remains and the old wood, were the traces of the ancient Tin-works. From Compstone onward to Holne, an elevated, heathy region, there occurred, at brief intervals, large tracts of ground which had been excavated to a shallow extent, and which have a very peculiar and easily recognisable appearance. They are of course covered with common vegetation, but the removal of the surface-earth, producing a depression of the level, has an effect so unique that the beholder, after he has had one or two of these " golfs," as they are termed, pointed out to him, readily detects one as he approaches it, and at once discriminates between these and any other irregularities in the country. These " golfs" are the spots where, in remote ages, the people searched for tin : the metal was found very near the surface, and was separated from the soil by the action of running water. A running stream (and this region abounds with such) was chosen, and so enclosed and directed as that its force should fall upon the metalliferous soil, when the lighter earth was washed away, and the heavier ore remained. This process was called *streaming ;* and it is believed that the ancient tinners were acquainted with no other mode of obtaining the metal.

Near Ringershots, the driver pointed out to us an ancient tin-stream, and it was with curious interest that I examined it. It is a romantic little gully, or glen, close by the road-side, all overhung and concealed by mountain-ashes, where a tiny thread of water now trickles with a metallic tinkle down the black, boggy soil. It has probably been unused for ages; but tradition has preserved the record of its former use, though the water which once made it available has long been drawn into other channels.

Thoughts of the old Phœnician rovers came up before my mind as the shadows of evening gradually shut out the scenery; and in imagination I followed the white metal, scarcely less precious than silver, across the stormy Bay of Biscay, through the Pillars of Hercules, and up the Mediterranean, till I saw it spread out in the market of Tyre; and, amidst the concourse of traffickers, heard the voices of the eager merchant-princes, crying,. as they strode in front of their stalls, " Bedil from Tarshish!*   Bedil!   Pure Bedil!   Bright Bedil!   Buy, buy, buy!" And then there came a twinkling of blue, and purple, and fine linen; and a chaffering and charming of many voices; and a Babel-hum of confused sounds.   But the familiar voice of the driver said, " Here we are at Holne Bridge;" and the vision vanished.   Yet I was glad that such was the last impression I had of Dartmoor.

* See the following Note.

APPENDIX.

ON THE TARSHISH OF SCRIPTURE.

(AN APPENDIX TO THE PRECEDING PAPER.)

THE question "What or where is Tarshish?" has been often asked, and often answered; whether satisfactorily is another matter. In the following notes I shall attempt, first, to collect in order the data which Scripture affords for the determination, and then see how far by these we may be able to identify the locality intended.

1. Tarshish was situated to the westward of Asia.

a. Because in Gen. x. 4, Tarshish is one of the descendants of Japheth, not of Ham, or Shem.

b. Because in Ps. xlviii. the ships of Tarshish are represented as threatening the coast of Palestine.

c. Because in Ps. lxxii. 10, Tarshish is associated with the isles,—the Hebrew expression for the maritime west.

d. Because in Isa. xxiii. merchants are represented as "passing over the sea," from Zidon, and from "the isle," (*i.e.* New Tyre), to Tarshish.

e. Because in Ezek. xxvii. 12, 25, "merchants of Tarshish," and "ships of Tarshish," are represented as crowding to the markets of Tyre.

2 D

f. Because Jonah (Jon. i. 3) found a ship at Joppa going to Tarshish.

2. Tarshish was a maritime, not an inland, country. (Sufficiently shewn by most of the passages above cited).

3. It was a place of maritime celebrity, at least as early as the times of David and Solomon (Ps. lxxii.).

4. It was a noted resort of the Tyrian traders. (See passages under prop. 1.)

5. It produced silver (Jer. x. 9 ; Ezek. xxvii. 12).

6. It produced iron (Ezek. xxvii. 12).

7. It produced tin (*Ibid.*).

8. It produced lead (*Ibid.*).

9. Tarshish continues to exist to the end of this dispensation. (See Ps. xlviii. ; Isa. ii. 16 ; Ps. lxxii. 10 ; all of which refer to the introduction or the course of the millennium).*

10. Tarshish is a maritime power at the end (Ps. xlviii. ; Isa. ii.).

11. The ships of Tarshish will be the chief medium for conveying the people of Israel "from far" to their own land, for millennial blessing (Isa. lx. 9 ; where note that Tarshish is the first of "the isles," or maritime countries).

12. Tarshish is an eminently mercantile people at the end (Ezek. xxxviii. 13).

13. Tarshish is a warlike people at the end (*Ibid.*). In this invasion of the land of Israel by Gog, the Prince of Ros, Mosc, and Tobl, the "young lions" of Tarshish, as well as its merchants, remonstrate against the invasion, though they do not take up arms against it. This indicates a commanding power.

Such are the data which Scripture gives us for the determination of the question ; a question which the prominent part to be acted by this power in events now surely at the doors, renders by no means unimportant.

An objection may occur against proposition 1, from 1 Kings x. 22 ; 2 Chron. ix. 21 ; where the navy of Solomon is repre-

* Some of these arguments will be cogent only to those who hold pre-millennial views of Divine Prophecy.

sented as bringing from Tarshish gold and silver, ivory, apes, and peacocks. These are products of Africa or southern Asia, rather than of Europe. If we could be quite certain that "peacocks" is the right rendering of תוכיים (*Tukiim*), it would fix this voyage to oriental Asia; and the same conclusion would be confirmed if we could certainly identify this fleet with that of Solomon and Hiram, which (1 Kings ix. 26–28) sailed from the Red Sea to Ophir. In 2 Chron. xx. 36, ships are represented as built at Ezion-gaber, (in the Red Sea,) for a voyage to Tarshish; but in the parallel passage (1 Kings xxii. 48) these ships are called "ships *of* Tharshish," intended "to go to *Ophir*."

The evidence in favour of a western Tarshish is far too strong to be set aside by these passages, allowing them the full weight of an opposing character, of which they are capable. At the utmost, one could only admit two Tarshishes, one oriental, the other occidental; and some have adopted this hypothesis. But to me it seems far more probable that, owing to the celebrity of the proper western Tarshish, the word came to be used indefinitely for any remote country beyond sea; and that the Hebrews used the expression, "a voyage to Tarshish," or "ships of Tarshish," as equivalent to "a voyage to foreign parts," or "ships that go foreign."

I come now to examine what light we have for identifying the true (*i.e.* the western) Tarshish, as characterized by propositions 1–13, with any country or town known in ancient or modern times, or in both.

Commentators have generally been content to fix on Tartessus, a city said to have been early founded by the Phœnicians, in the south of Spain. There is much doubt about the actual position of this city: according to Herodotus (iv. 152), it was outside the straits of Gibraltar; but all agree that it had ceased to exist before the Christian era. This fact alone is fatal to the identification of Tarshish with Tartessus.* Whatever points of agreement there may have been, Tartes-

* I think it probable that this Phœnician colony was named after the original Tarshish.

sus entirely fails to meet the requirements of propositions 9–13. Tartessus has long disappeared; and even if we were to extend the identification to the region which once bore that city, we should fare no better. Spain has neither maritime nor mercantile eminence ; her ships are not likely to be the chief agency used in the transport of returning Israel ; nor does she possess any " young lions " likely to remonstrate with an invading despot.

Britain meets every requirement, without a single valid objection.

1. Britain is west of Asia, peopled by Japheth's descendants.
2. Is a maritime country.
3. Was a place of maritime celebrity in very early ages ; since—
4. The Phœnicians resorted to Cornwall (Strabo iii. 175) in very remote ages ; and specially for the metals, tin, and lead. Mines yet exist in Cornwall, and mountains of slag, the evidence of works which date far beyond the historic era.
5. Cornwall produces silver.
6. It produces iron.
7. It produces tin. This was the great staple commodity of Cornwall from time immemorial (as it still is); and gave to the region the name of Κασσιτερίδες (*Cassiterides*), or "the tin coasts."
8. It produces lead.
9. Britain still exists.
10. Is now a maritime power.
11. Has been much honoured of God, and is more likely than any other power to be the medium for conveying the Jews to Palestine (Isa. lx. 9). Britain is at *the head* of maritime nations.
12. Britain is eminently a nation of merchants.
13. Her well-known bravery and love of liberty well suit the remonstrance of Ezek. xxxviii. 13 ; while her pacific policy may account for her interference being *limited* to remonstrance.

Many collateral passages might be enumerated, which indicate Britain as destined to carry out God's purposes of mercy at the end, when judgment falls on the Papal earth. Enough, however, has been shewn to warrant the confident conclusion that the Tarshish of Scripture has always been, and never can be other than GREAT BRITAIN.

INDEX.

BALLANTYNE AND CO., PRINTERS, EDINBURGH.

Printed in the United States
By Bookmasters